Shengnan Zou

Magnetization of High Temperature Superconducting Trapped-Field Magnets

Karlsruher Schriftenreihe zur Supraleitung Band 019

HERAUSGEBER

Prof. Dr.-Ing. M. Noe
Prof. Dr. rer. nat. M. Siegel

Eine Übersicht über alle bisher in dieser Schriftenreihe
erschienene Bände finden Sie am Ende des Buches.

Magnetization of High Temperature Superconducting Trapped-Field Magnets

by
Shengnan Zou

KIT Scientific Publishing

Karlsruher Institut für Technologie
Institut für Technische Physik

Magnetization of High Temperature Superconducting
Trapped-Field Magnets

Zur Erlangung des akademischen Grades eines Doktor-Ingenieurs
von der Fakultät für Elektrotechnik und Informationstechnik des
Karlsruher Instituts für Technologie (KIT) genehmigte Dissertation

von Shengnan Zou aus Jilin, China

Tag der mündlichen Prüfung: 19. Juli 2017
Referenten: Priv.-Doz. Dr. Francesco Grilli
Prof. Frédéric Sirois, B.Ing., Ph.D.
Prof. Dr.-Ing. Mathias Noe

Impressum

KIT Scientific Publishing

Karlsruher Institut für Technologie (KIT)
KIT Scientific Publishing
Straße am Forum 2
D-76131 Karlsruhe

KIT Scientific Publishing is a registered trademark
of Karlsruhe Institute of Technology.
Reprint using the book cover is not allowed.

www.ksp.kit.edu

Print on Demand 2017 – Gedruckt auf FSC-zertifiziertem Papier

ISSN 1869-1765
ISBN 978-3-7315-0715-4
DOI 10.5445/KSP/1000073433

Zusammenfassung

Sowohl massive Hochtemperatur-Supraleiter (HTS) als auch gestapelte Hochtemperatur-Supraleiter-Bandleiter können als Permanentmagnete verwendet werden, die deutlich höhere Magnetfelder als konventionelle Permanentmagnete bereitstellen. Dies erreichen sie durch den Einschluss des magnetischen Feldes im supraleitenden Zustand. Diese Supraleiter mit eingeschlossenem Magnetfeld (trapped-field magnets) sind vielversprechend für eine Vielzahl von elektrischen Anwendungen, die permanente Magnete verwenden. Die gepulste Feldmagnetisierungs- (PFM) Methode, zur Magnetisierung der Supraleiter wird dabei immer interessanter, da sie kosteneffizient, kompakt und flexibel im Einbauzustand angewendet werden kann. Dies ist entscheidend für die Realisierung von praktischen Anwendungen von trapped-field magnets. Die größte Herausforderung der PFM ist, dass die generierten Magnetfelder bisher noch schwächer sind als die durch Feldkühlungs- oder Nullfeldkühlungsmethoden erzeugten Magnetfelder. Dieser Unterschied entsteht durch den Temperaturanstieg bei schnellen Flussänderungen. Die Motivation dieser Arbeit ist es, den dynamisch magnetischen Fluss während der PFM zu untersuchen, zu verstehen und mögliche Strategien zur Verbesserung des trapped fields, produziert durch die PFM, vorzuschlagen.

Für HTS Massivmaterialien werden in dieser Arbeit 2D elektromagnetisch-thermisch gekoppelte Modelle genutzt, um den PFM Prozess zu simulieren. Außerdem werden wichtige Experimente aus der Literatur überprüft. Die Untersuchungen ergeben, dass die Simulationsergebnisse stark parameterabhängig sind und die 2D Modelle, aufgrund der Inhomogenität des HTS Massivmaterials, das Verhalten nicht genau beschreiben können.

HTS Bandleiterstapel werden mit neuartig entwickelten 2D und 3D Modellen, basierend auf der Finiten Elemente Methode, simuliert. Anhand von 2D elektromagnetisch-thermisch gekoppelter Modellierung werden zwei Strategien zur Verbesserung des trapped fields, welches durch die PFM produziert wird, vorgeschlagen. Eine der beiden Methoden verwendet eine Spule zur Flussdichteverteilung. Dieser magnetisiert den Stack. Bei dem anderen wird eine optimale Pulssequenz mit einem Anfangspuls, der groß genug ist und aufeinanderfolgenden Pulsen von absteigenden Amplituden mit infinitesimalen Intervallen verwendet. Die zweite Strategie wurde qualitativ durch Experimente validiert. Anhand von 3D Modellierung wird ein quadratisch, flacher Stack und ein gekrümmter Stack untersucht. Der gekrümmte Stack ist aufgrund seiner geometrischen Anwendbarkeit für elektrische Maschinen von Interesse. Die magnetische Feldverteilung von flachen und gekrümmten Stacks, die durch Feldkühlung magnetisiert werden, wird mit 3D statischen Modellen kalkuliert und die Ergebnisse stimmen quantitativ mit experimentellen Messungen überein. Bei Verwendung von 2D und 3D elektromagnetisch-thermisch gekoppelten Modellen werden die durch PFM magnetisierten flachen und gekrümmten Stacks simuliert. Zur Beschleunigung der Simulationen wird während der Gittererstellung auf eine gleiche Gittergröße geachtet. 2D und 3D Modelle liefern für die Berechnung des trapped fields für einen flachen, quadratischen Stack nahezu ähnliche Ergebnisse. Diese Ergebnisse stimmen quantitativ mit den in der Literatur angezeigten Ergebnissen überein. Die gekrümmten und flachen Stacks, die durch die PFM magnetisiert wurden, zeigen ein ähnliches Verhalten. Die magnetische Feldverteilung von einem gekrümmten Stack verändert sich durch geometrische Deformation, was qualitativ durch Experimente bestätigt wurde.

Diese Arbeit repräsentiert den Stand der Technik über die Simulation für HTS beschichtete Leiterstacks, die magnetisiert wurden. Zum ersten Mal wurde deren elektromagnetisch und thermisches Verhalten in 3D und ohne vereinfachte Annahmen der geometrischen Struktur in 2D simuliert.

Abstract

High temperature superconducting (HTS) bulks and stacks of coated conductors can be magnetized to become trapped-field magnets that provide much stronger magnetic fields than those reachable with conventional permanent magnets. The trapped-field magnets are promising for a variety of electrical applications that use permanent magnets. The pulsed field magnetization (PFM) method is attracting attention as it can provide cost-effective, compact and flexible *in situ* magnetization, which is critical for realizing practical applications of trapped-field magnets. The key challenge of PFM is that the produced trapped fields are generally lower than those produced by field cooling or zero field cooling methods due to the temperature increase caused by fast flux motions. The motivation of this work is to investigate and understand the flux dynamics during PFM and to propose possible strategies to improve the trapped field produced by PFM.

For HTS bulks, 2D electromagnetic-thermal coupled models are used to simulate the PFM process and relevant experiments in the literature are reviewed. The findings are that the simulation results are highly parameter dependent and 2D models cannot properly describe HTS bulks due to inhomogeneity of HTS bulks.

Stacks of HTS coated conductors are simulated with newly developed 2D and 3D models based on the finite element method. By means of 2D electromagnetic-thermal coupled modelling, two strategies are proposed to improve the trapped field produced by PFM. One is to use controlled magnetic density distribution coils to magnetize the stack. The other is to apply an optimal pulse sequence with a large enough initial pulse and successive pulses of descending amplitudes with infinitesimal intervals. The second strategy is qualitatively validated by experiments. With 3D modelling, a square flat stack and a curved stack are investigated. The curved stack is of interest because of its geometrical applicability for electrical machines. The magnetic field distributions of flat and curved stacks magnetized by field cooling are calculated with 3D static models and the results quantitatively agree with experimental measurements. Using 2D and 3D electromagnetic-thermal coupled models, the flat and curved stacks magnetized by PFM are simulated. Homogenization and mesh techniques are used to speed-up the simulations. It is found that 3D and 2D models obtain close trapped fields for a flat square stack and the results quantitatively agree with experiments reported in the literature. The curved and flat stacks magnetized by PFM show similar behaviors. The magnetic field distribution of a curved stack is changed due to geometrical deformation, which is qualitatively validated by experiments.

The numerical models developed and used in this thesis work constitute the state-of-the-art simulation tools for investigating the magnetization of HTS coated conductor stacks: for the first time, the electromagnetic and thermal behavior of such stacks was simulated in 3D, and without simplifying assumptions on the geometrical structure in 2D.

Contents

List of Figures

List of Tables

Abbreviations

CC	Coated conductor
CMDCs	Controlled magnetic density distribution coils
FC	Field cooling
FeSC	Iron-based superconductor
FEM	Finite element method
GdBCO	Gadolinium barium copper oxide ($GdBa_2Cu_3O_{7-\delta}$)
HTS	High temperature superconductor
IMRA	Iteratively magnetizing pulsed field method with reducing amplitude
LTS	Low temperature superconductor
MgB_2	Magnesium diboride
MPSC	Multi-pulse technique with step-wise cooling
PFM	Pulsed field magnetization
ReBCO	Rare-earth barium copper oxide materials
SPA	Successive pulsed-field application
YBCO	Yttrium barium copper oxide ($YBa_2Cu_3O_{7-\delta}$)
ZFC	Zero field cooling

Acknowledgements

First and foremost, I would thank my supervisor Dr. Francesco Grilli, who provided me invaluable instructions, discussions and support on work and life at all times. It is great fortune for me to work with him. I would also thank Prof. Mathias Noe, who gave advices and helped me to proceed in a timely manner. I would thank Dr. Victor M. R. Zermeño as he offered so much precious technical help.

The experiments in terms of pulsed field magnetization were carried out with the generous support of Dr. Anup Patel, Algirdas Baskys and Prof. Bartek A. Glowacki at the University of Cambridge, who I cannot appreciate enough. The AC loss measurement was supported by Aurelien Godfrin, Dr. Anna Kario and many colleagues. My visit to Shanghai Jiaotong University was kindly hosted by Dr. Zhiyong Hong, Dr. Zhuyong Li and all the students.

I can never forget all the students, colleagues and friends in the institute, with whom we shared all the happy time at and after work and the delicious sweets. I would thank all my Chinese fellows who accompanied me experiencing happiness and difficulties during the years away from our homeland.

I owe everything to parents Min Qi and Shuzhong Zou. They have always been supporting me with no regret and no reservation. I feel sorry that I have been away for the years and could not be accompanying and rewarding them. I can do nothing but express my gratitude.

I would give my final thank with love to my husband Ling Zhang. He respected my choice to start the PhD, have always been supporting me during the years, and proposed to me in Germany. We can finally reunite in China after three years of separation with the completion of the thesis.

Karlsruhe, May 2017

Shengnan Zou

1 Introduction

As members of the type-II superconductor family, high temperature superconductor (HTS) bulks or stacks of coated conductors (CCs) can work as permanent magnets, namely trapped-field magnets, to provide extremely strong magnetic fields by trapping persistent currents. The current remarkable world record trapped field is 17.6 T provided by a GdBCO bulk at 26 K [1], much higher than the fields provided by conventional permanent magnets, which are typically lower than 2 T.

Towards the practical application of such magnets, one of the key issues is to provide *in situ* magnetization methods. The traditional magnetization methods, including field cooling and zero field cooling, need to create a quasi-static background field of a quasi-static background field of at least the same or twice the magnitude of the target trapped field, respectively. Large magnetization coils, strong power supplies and other supplementary equipment are required; as a result, the magnetization is time consuming and expensive and the set-up is huge. The pulsed field magnetization (PFM), which is characterized by compactness, mobility and cost advantage, is a promising method for practical applications. However, a key challenge is that the trapped fields produced by PFM are much lower than those produced by field cooling method, because the temperature increases greatly during PFM due to fast flux motion.

This thesis is devoted to investigate the flux dynamics during PFM and to provide strategies to improve the trapped field produced by PFM by means of numerical simulations with electromagnetic-thermal coupled models in both 2D and 3D. Experiments are carried out for validating the numerical simulations.

The main objects of study of this thesis are stacks of coated conductors, which have been attracting interest in recent years and regarded as alternatives to bulks. Benefits of stacks of coated conductors include improved cost-effectiveness, more flexible geometries and high mechanical strength. Due to the extensive investigations HTS bulks as reviewed in [2] and the relatively simple geometry and structure of bulks, the thesis starts with simulations of HTS bulks. The goal is to shed light on several basic aspects of the numerical modelling of the flux dynamics during PFM. The major contributions of this thesis are the investigation of flux dynamics in stacks of coated conductors during PFM and the proposal of possible strategies to improve the produced trapped field. The thesis is organized in chapters. The contents are listed below.

Chapter 2 provides a brief introduction to high temperature superconductors. Relevant properties of HTS are introduced, including the temperature- and magnetic field-dependent properties, electromagnetic behaviors and the origins of the AC loss.

Chapter 3 introduces the numerical modelling of superconductors. Analytical and numerical models are briefly reviewed. The implementation of the ***H***-formulation of Maxwell's equations in the numerical simulations of the thesis is described in detail. The heat transfer equation and its coupling with the ***H***-formulation are described as the modelling framework of the thesis.

Chapter 4 contains simulations of HTS bulks with 2D axisymmetric electromagnetic-thermal models. Based on the results in the literature, HTS bulks magnetized by PFM with controlled magnetic density distribution coils are simulated. The results show some discrepancy with literature; accordingly, a discussion about the influence of parameters on the simulation is made. Finally, experimental evidence is reviewed for benchmark of the simulations of bulks.

In Chapter 5, the PFM of HTS coated conductor stacks is investigated. First, simulations with 2D electromagnetic-thermal coupled models are carried out with realistic structures and comprehensive material properties of coated conductors. Based on the model, two strategies are investigated to improve the trapped fields produced by PFM: one is to use the new arrangements of controlled magnetic density distribution coils; the other is to optimize the pulse sequence used for PFM. Then, the 2D modelling is extended to 3D. A curved stack is investigated for its geometrical adaptability in the rotor of an electrical machine. A common square flat stack and a curved stack magnetized by field cooling and PFM are simulated and compared. The methods to reduce the computation time are discussed. Finally, experiments are carried out for validating the strategy of PFM with multi-pulse technique proposed by the 2D simulations. Flat stacks and curved stacks are magnetized by field cooling and PFM; and the results are compared with the 3D simulations.

Finally, Chapter 7 summarizes the contributions of this thesis work and suggests perspectives in the research on trapped-field magnets magnetized by PFM.

2 Basics of Superconductivity

This chapter introduces the basics of superconductivity. Specifically, properties of type-II superconductors including the critical parameters and voltage-current dependence that will be used in later simulations are described. Losses in superconductors are described, especially hysteresis and eddy current losses. HTS bulk and coated conductor (CC) materials, which constitute two types of trapped-field magnets (TFMs), are introduced. Finally, applications and magnetization methods of trapped-field magnets are described to propose the motivation of the thesis work.

2.1 Introduction to Superconductivity

Superconductivity was first discovered in mercury by Heike Kamerlingh Onnes in 1911 [3], [4], [5]. So far, many elements and compounds have been found to be superconductors. Type-II superconductors which present "non-ideal" superconductivity are the superconductors for practical applications.

2.1.1 Basic Properties

The essential characteristics of superconductors are zero electrical resistance and the Meissner effect. The former is obvious since the discovery of superconductivity. Below a certain temperature, namely the critical temperature T_c, a superconductor has no electrical resistivity when carrying constant currents. This was demonstrated by observing no current decays in a close superconducting ring [4].

The Meissner effect was discovered over 20 years after the discovery of superconductivity [6]. A superconductor excludes magnetic fields from its interior by generating a screening current on the surface to oppose external magnetic fields. This happens regardless of whether the external fields are applied before or after the material becomes superconducting. The Meissner effect distinguishes a superconductor from a perfect conductor, because the absolute field expulsion cannot be derived from the Maxwell's equations and zero electrical resistivity [4].

2.1.2 Type-I and Type-II Superconductors

Depending on the magnetic property, superconductors can be categorized into two types: type-I and type-II superconductors. A superconductor restores its resistivity when it is exposed to a sufficiently large magnetic field. For type-I superconductors, the Meissner state is also lost when the strength of the external magnetic field reaches this magnitude (the intermediate state due to geometry factors is not discussed here). This field which destroys superconductivity completely is known as the critical field B_c. The critical field depends on the temperature T, and the dependence is approximately parabolic as shown in Fig. 2.1(a). Almost all pure elemental superconductors are type-I. The critical fields for type-I superconductors were found to be generally lower than 120 mT; as a result, type-I superconductors cannot carry large currents and are not suitable for practical applications [4].

Type-II superconductors exhibit a complex behavior in magnetic fields, as shown in Fig. 2.1(b). Below a certain magnetic field, namely the lower critical field B_{c1}, the material exhibits the Meissner state and external fields are completely excluded from its interior. When the applied field is above B_{c1}, normal zones start to appear as vortices of quantized magnetic fluxons [7], [8]. Only when the applied field is above the upper critical field B_{c2}, the material reverts to the normal state completely. While the external field is between B_{c1} and B_{c2}, a type-II superconductor is in the so-called mixed state in which normal cores exist. Alloys and complex oxide ceramics superconductors are type-II.

Type-II superconductors, especially those that have defects acting as "pinning centers" in the crystal structures, are ideal for practical applications. When a type-II superconductor carries a current, the electrons experience Lorentz forces due to the existence of magnetic flux through normal cores. This force can cause the normal cores to move; as a result, the currents go through normal zones and resistivity appears. However, the existence of defects can act as pinning centers of magnetic fluxes. They produce pinning forces to impede magnetic fluxes from moving. As a result, type-II superconductors with strong flux pinning effects can carry high currents without energy dissipation. Such superconductors are also called "hard" type-II superconductors. At present, impurity is introduced on purpose to improve the current-carrying capacity of superconductors.

The hard type-II superconductors are the research subjects of the thesis, specifically, high temperature superconductors made of rare-earth barium copper oxide (Chapter 2.2). Macroscopically, current-carrying capacity of practical type-II superconductors can be described by the critical current density J_c, which depends on the magnetic field and the temperature (Chapter 2.3). The electromagnetic behaviors can be described by a simplified phenomenological model or an extended voltage-current equation (Chapter 2.4).

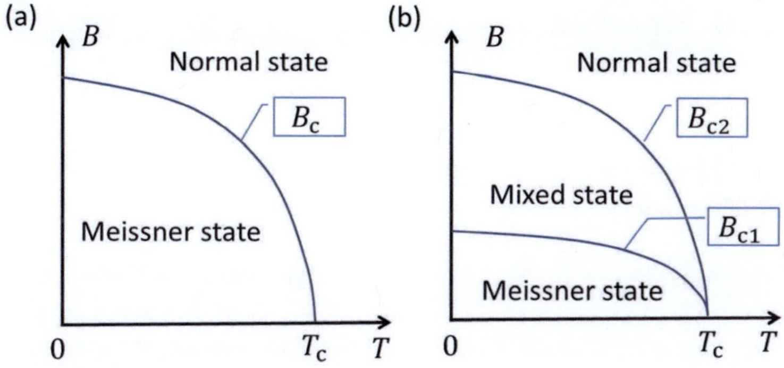

Figure 2.1: Temperature dependence of the critical fields of (a) type-I and (b) type-II superconductors, respectively.

2.2 High Temperature Superconductors

In 1986, a new family of superconductors, oxocuprates, was discovered [9]. Many of the oxocuprates have the critical temperatures T_c above the boiling point of liquid nitrogen (77 K), for example, $YBa_2Cu_3O_{7-\delta}$ ($T_c \approx 92$ K), $Bi_2Sr_2Ca_2Cu_3O_{10-\delta}$ ($T_c \approx 110$ K) [10]. Due to their relatively high critical temperatures, these superconductors are also called high temperature superconductors (HTS) in contrast to low temperature superconductors (LTS). High temperature superconductors can use liquid nitrogen as coolant, which is much cheaper and easier to handle than liquid helium.

In the first decade of the 21st century, another two families of superconductors with relatively high critical temperatures were discovered: magnesium diboride (MgB_2) with T_c of 39 K [11] and iron-based superconductors (FeSC) with T_c up to 58 K [12], [13], [14].

This thesis focuses on trapped-field magnets made of rare-earth barium copper oxide materials (ReBCO). ReBCO materials are produced in two forms: bulk material or coated conductor (CC). The former can be magnetized to be a magnet; the latter, which appears as long conducting tapes, can be stacked to become a composite bulk [2].

2.2.1 Bulk Material

Most HTS bulks are made of ReBCO [2]. Currently MgB_2 and FeSC bulks can also be made into bulks [15], [16], [17], [18]. One common example of ReBCO is $YBa_2Cu_3O_{7-\delta}$ (YBCO or Y-123). Yttrium can be replaced by other rare elements (La, Tl, Sm or Gd) and there are also other compositions.

ReBCO bulks are produced by melt processing techniques [19]. Pressed mixed powders depending on targeted compositions are placed together with a small single crystal into an oven and processed with heat treatment. In order to improve the quality and speed of production, multiple seeds may be placed [20]. As a result, HTS bulks may be rather inhomogeneous following growth sector boundaries [21]. Moreover, HTS bulks are fragile ceramics and may crack when experiencing stresses during magnetization, so extra mechanical reinforcement is required [22].

A bulk is usually produced to be a cylinder or pellet. A ring-shape bulk can also be produced for particular needs [23]. At the moment, J_c of a bulk is in the range of 10^8 to 10^9 A/m^2 with no external field at 77 K. It is not straightforward to characterize an HTS bulk. One way is to take a slice from the material and measure the critical current by applying a transport current directly. Since a bulk is usually inhomogeneous, it is not easy to evaluate J_c of the entire bulk [21]. Another way is to map the magnetic fields around the bulk in or after application of an external magnetic field and calculate the current density in the bulk by inversion operation [24], [25].

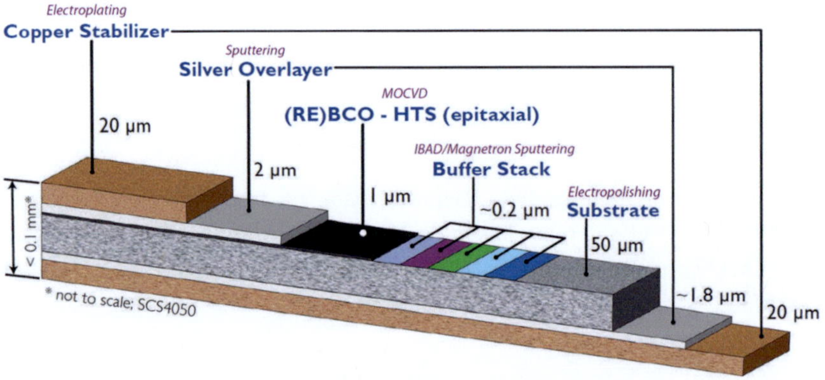

Figure 2.2: Configuration of a SuperPower ReBCO HTS tape [26], [27].

2.2.2 Coated Conductor

ReBCO material can be also produced by film deposition. In this way, the material is manifest as small films or long wires. The former are used for electronic devices; the latter are used for transporting strong currents, for example, cables. An HTS wire is usually a flat tape that is hundreds of micrometers in thickness and millimeters in widths.

A ReBCO HTS tape is also called a coated conductor (CC) because it is fabricated with layer-by-layer film coating as shown in Fig. 2.2 [26], [27]. The ReBCO layer is usually deposited on a metallic substrate with a very thin stack of buffer layers that acts as a biaxial texture template [27]. The superconducting layer produced by chemical or physical deposition is usually only around 1 μm thick. Extra layers of metal may be added to provide stabilization according to different applications. The critical current density J_c of the HTS layer is in the order of 10^{10} A/m^2 at 77 K, while the engineering critical current density $J_{c,e}$, which is the critical current of an HTS tape divided by the entire cross-section area of the tape, is in the order of 10^8 A/m^2.

ReBCO tapes are more mechanically robust compared to bulks due to the presence of metallic substrates. However, there is a minimum bending diameter that the tape can withstand. When a tape is bent over a diameter smaller than the minimum bending diameter, the critical current can be degraded. The minimum bending diameter of a commercial tape is currently around 20 mm [26].

2.3 The Critical Parameters

Superconductivity only exists when the temperature is below the critical temperature. Besides, a superconductor will also restore resistivity when the magnetic field exceeds its critical field (B_{c2} for type-II) or the current density exceeds the critical current density J_c. The critical temperature, the critical field and the critical current density depend on each other and compose a critical surface $J_c(B,T)$ as shown in Fig. 2.3.

When implemented in numerical models, the critical current density J_c as a function of the magnetic field and temperature can be expressed by direct interpolations or with simplified equations based on experiments. One widely used empirical equation of magnetic field dependence of J_c for modelling type-II superconductors, known as the Kim model [2], [28], [29], is

$$J_c(B) = \frac{J_{c0}}{1 + B/B_0}, \tag{2.1}$$

where J_{c0} and B_0 are temperature-dependent parameters. This equation fits generally for most ReBCO bulks without abnormal "fish-tail" or "peak" effects [30], [31]. The temperature dependence of J_{c0} can be expressed as [2]

$$J_{c0}(T) = \alpha\left[\left(1 - \frac{T}{T_c}\right)^2\right]^{1.5}, \tag{2.2}$$

where T_c is the critical temperature and α is a parameter in units of A/m^2.

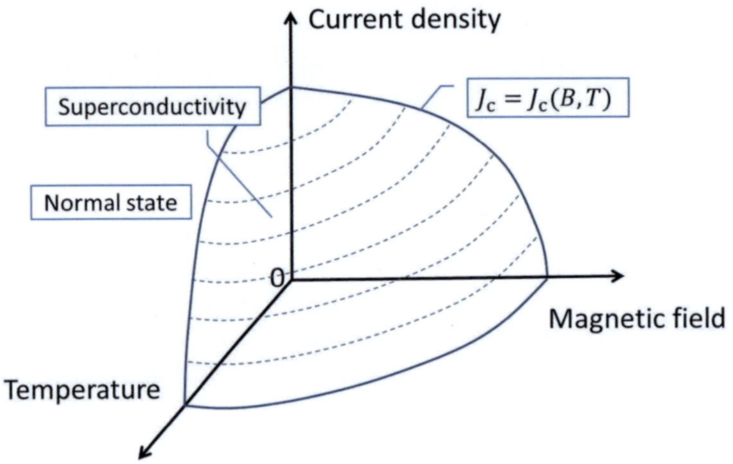

Figure 2.3: The critical surface for a type-II superconductor, which is the boundary between the superconducting and the normal states. Superconductivity exists when the magnetic field B, the temperature T and the current density J are below the critical surface.

All the above expressions do not take into account the orientation of the magnetic field with respect to the superconductor. ReBCO tapes, however, exhibit obvious anisotropic behaviors in magnetic fields. For simplification, the critical current density J_c that depends not only on the magnitude of the magnetic field but also on the orientation can be expressed as [32], [33]

$$J_c(B_\perp, B_\parallel) = J_{c0}(T)\left(1 + \frac{\sqrt{B_\perp^2 + k^2 B_\parallel^2}}{B_0}\right)^{-\beta}, \tag{2.3}$$

where $B_\perp$ and $B_\parallel$ are the components of the magnetic field perpendicular and parallel to the flat surface of the tape, respectively, k is the anisotropy parameter ($k < 1$) and the rapidity of the J_c reduction with the magnetic field is given by B_0 and β. The parameters

k, B_0 and β are temperature dependent. For many commercial tapes, the angular dependence of the critical current density on the magnetic field can be very complicated, due to artificial pinning centers aiming at improving the in-field behavior of the superconductors [34], [35].

2.4 Voltage-Current Dependence

The electromagnetic behavior of an ideal hard type-II superconductor, characterized by the presence of vortices in its mixed state [7], [8], can be described by the critical state model (CSM), which represents a macroscopic average of flux vortices. Furthermore, the *E-J* power law can be used to further account for flux creep or flow behavior of a non-ideal hard superconductor [36], [37].

2.4.1 The Critical State Model

First proposed in 1962 to describe a hard type-II superconductor in response to an external magnetic field, the basic premise of the critical state model is: "there exists a limiting macroscopic superconducting current density that a hard superconductor can carry; and further, that any electromotive force, however small, will induce this full current to flow into an inhomogeneous mixed-state superconductor" [38], [39].

The critical state model accounts for quasi-static conditions that flux vortices are firmly pinned and unaltered unless external field is changed. The flux vortices move to reach a new equilibrium state instantly when the external field changes. The Lorentz force has to overcome the pinning force to move the flux vortices and this process generates dissipation. In its original formulation, the critical state model assumes that the temperature is unchanged and neglects the magnetic field dependence of the critical current density J_c.

The critical state model makes the following phenomenological assumptions: a) the current density can only be J_c (critical region) or 0 (flux free region); b) with the application of an external field, the critical region always appears from the border of the superconductor and penetrates to the interior; c) the distribution of current density always tends to prevent the magnetic flux density in the interior of the superconductor from changing.

A classic example for critical state model is the description of the magnetization process of an infinite slab of thickness $d/2$ in a magnetic field parallel to its flat surface as shown in Fig. 2.4. Figure 2.4(a) and (b) presents the zero field cooling (ZFC), which means that the material becomes superconducting without the presence of an external field.

Figure 2.4(c) presents the field cooling (FC), which means that the material becomes superconducting at the presence of an external field. Following Ampere's law, the slope of the magnetic field in the slab equals J_c. There is a penetration field $H_p = J_c d/2$ at which the two current fronts meet and there is no flux-free region.

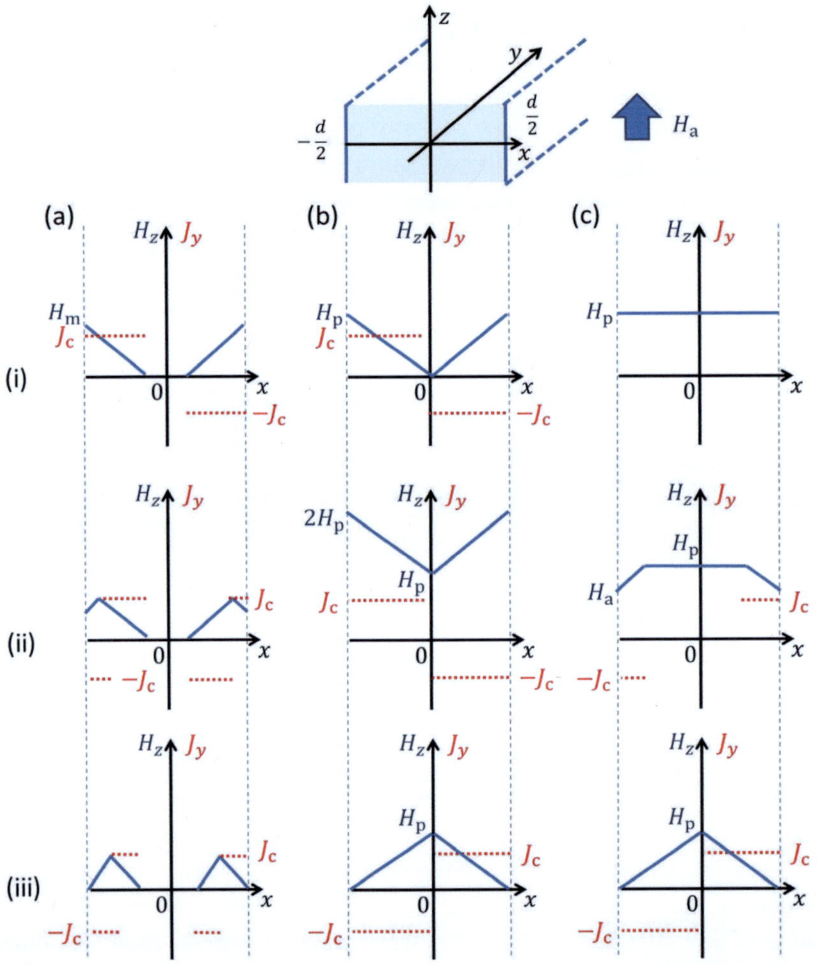

Figure 2.4: The magnetic field and current density distribution of an infinite HTS slab described by the critical state model. (a) An external magnetic field of amplitude $H_m < H_p$ is applied to the slab and then retrieved (zero field cooling, ZFC); (b) ZFC with the external field of amplitude $2H_p$; (c) the slab turns superconducting in an external magnetic field of magnitude H_p and the magnetic field is retrieved (field cooling, FC).

For zero field cooling, when an external field smaller than H_p is applied as shown in Fig. 2.4(a), currents penetrate from the surface of the slab. When the external field decreases to zero, negative currents start to appear from the surface. Finally, there remain double loops of currents in opposite directions and there is a flux-free region. In this case, the slab is not fully magnetized.

As shown in Fig. 2.4(b), if an external field of magnitude $2H_\mathrm{p}$ is applied, currents penetrate from the surface of the slab until the slab is fully penetrated. Then, the current distribution remains the same when the external field increases from H_p to $2H_\mathrm{p}$. When the external field is decreased, negative currents start to appear from the surface. Finally, there remains a single-loop current and the central field of the slab is H_p. In this case, the slab is fully magnetized.

For field cooling, the slab is in an external field larger than H_p and then becomes superconducting. When the external field is reduced, currents appear from the surface to maintain the flux in the center of the slab. Finally, a single-loop current remains with the central field equal to H_p.

Both zero field cooling and field cooling leave a remnant current and field in the slab, which means that the magnetization of the slab is hysteretic. The slab traps a field to become a permanent magnet. The magnetization of a hard type-II superconductor is irreversible, similar to a ferromagnetic material [40]. The local magnetization is the field created by the currents. Defining M as the volume average of the local magnetization, the magnetization loops of an infinite slab by magnetic field H_a of different amplitudes H_m based on the critical state model are shown in Fig. 2.5. The magnetization is normalized by the saturation magnetization ($M_\mathrm{sat} = J_\mathrm{c}\, d/4$); the applied field H_a and peak amplitude H_m is normalized to the penetration field H_p.

Since the applied field is parallel to the flat surface of the infinite slab, there is no demagnetization field and initially $M = -H$. For $H_\mathrm{m}/H_\mathrm{p} < 1$, the magnetization does not saturate and there is a flux-free region as shown in Fig. 2.4(a). For $H_\mathrm{m}/H_\mathrm{p} \geq 1$, the magnetization saturates when the applied field reaches the penetration field, when the slab is fully penetrated and no flux-free region exists. For $H_\mathrm{m}/H_\mathrm{p} \geq 2$, the remnant magnetization (when the applied field is zero) reaches the maximum possible value (saturation magnetization), which corresponds to the zero field cooling situation in Fig. 2.4(b)-(iii).

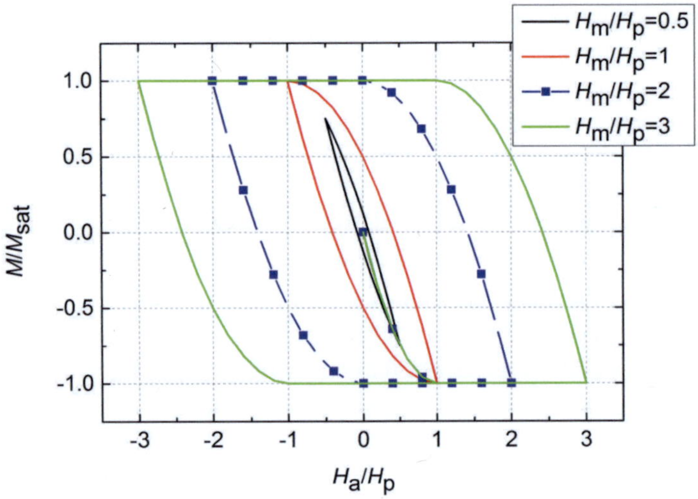

Figure 2.5: The magnetization loops for an infinite slab by applied fields of different amplitudes (H_m). The magnetization is normalized by the saturation magnetization ($M_\mathrm{sat} = J_\mathrm{c}d/4$); the applied field ($H_\mathrm{a}$) and peak amplitude ($H_\mathrm{m}$) is normalized to the penetration field ($H_\mathrm{p} = J_\mathrm{c}d/2$).

Superconductors of finite geometries behave in a similar way: currents always start to appear and change from the boundaries and the superconductor traps a remnant field finally. However, the field and current distribution are much more complicated and, generally, cannot be solved for simple expressions analytically. Such cases can be solved numerically.

2.4.2 *E-J* Relationship

The critical state model gives an intuitive picture of the behavior of HTS, but flux creep and flux flow cannot be taken into account [36], [37]. Based on the experimental results on the voltage-current measurements, the magnitudes of the electrical field E and the current density J is expressed as the *E-J* power law [33], [41]

$$E = E_\mathrm{c}\left(\frac{J}{J_\mathrm{c}}\right)^\mathrm{n}, \qquad \textbf{(2.4)}$$

where E_c is the characteristic electric field. For HTS applications, E_c is usually set to be 10^{-4} V/m according to the international standard [42]; the exponent parameter n describes the steepness of the *E-J* curve. The critical current density J_c can be temperature and magnetic field dependent. The non-linear resistivity ρ_{p} derived from the *E-J* power law can be expressed as

$$\rho_{\mathrm{pl}} = \frac{E_{\mathrm{c}}}{J_{\mathrm{c}}}\left(\frac{J}{J_{\mathrm{c}}}\right)^{\mathrm{n}-1}. \tag{2.5}$$

For $n = 1$, the equation represents a linear *E-J* dependence like normal metals. For the $n \to \infty$ limit, the equation corresponds to the critical state model as shown in Fig. 2.6.

When J is much larger than J_{c}, however, the power law is not proper because the resistivity continues to increase exponentially and even exceed the normal state resistivity.

There are no solid experiments to reveal the intrinsic voltage-current dependence of the HTS material in the over critical region. To give a reasonable slope of the transition region from the superconducting to normal state, a saturating resistivity can be added in parallel to the power law resistivity [43], [44], [45]

$$\rho_{\mathrm{SC}} = \frac{\rho_{\mathrm{pl}}\rho_{\mathrm{n}}}{\rho_{\mathrm{pl}} + \rho_{\mathrm{n}}}, \tag{2.6}$$

where ρ_{n} is the normal state resistivity. As shown in Fig. 2.7, the modified resistivity converges to the normal resistivity for large currents.

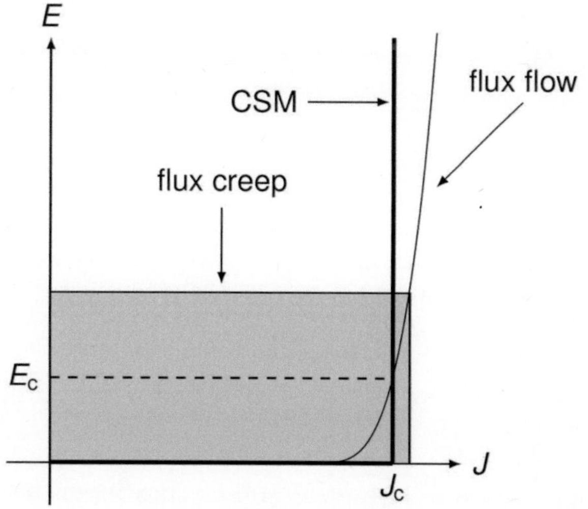

Figure 2.6: *E-J* characteristics of HTS for the critical state model (thick line, CSM) and the power law (thin line) [33].

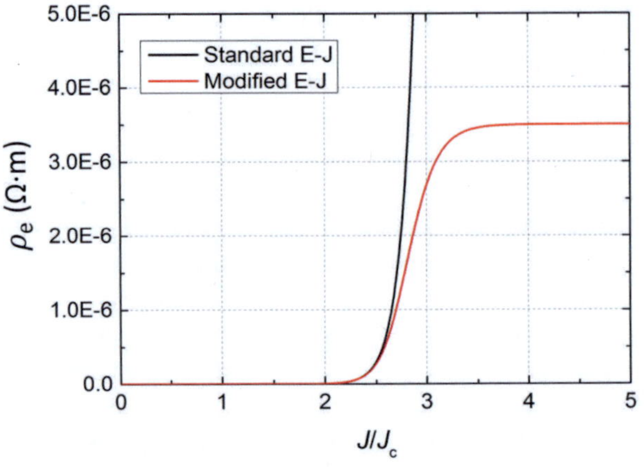

Figure 2.7: Modified electrical resistivity of HTS (Eq. 2.6) compared with a standard power law (Eq. 2.5) with the normalized current density J/J_c [45].

2.5 AC Loss in HTS

A superconductor is characterized with zero resistivity below the critical temperature. However, a type-II superconductor is hysteretic and dissipates energy whenever magnetic flux lines inside it moves. The movement of magnetic flux lines can happen when a superconductor is exposed to a varying magnetic field or carries a varying current. Such loss dissipated in devices with HTS components is often called AC loss, but the loss is not necessarily caused by AC excitations.

In practice, superconductors can be presented as tapes composed of normal metals or ferromagnetic substrates and may have complicated structures such as twisted filaments or multi layers. These supplementary parts or structures can also cause significant losses. There are four types of loss contributions in practical HTS tapes as schematically shown in Fig. 2.8 [33]:

1) Hysteresis loss: produced by movement of magnetic flux in the superconducting material.

2) Eddy current loss: produced by current in the normal metal induced by a varying magnetic field.

3) Coupling loss: caused by currents in normal metallic parts flowing between separate superconducting filaments.

4) Ferromagnetic loss: caused by hysteresis loops in magnetic parts if present.

For the pulsed field magnetization of trapped-field magnets in this thesis, the stacks and bulks have no magnetic parts or filaments. The hysteresis loss and the eddy current loss are the only relevant kinds and are introduced below.

2.5.1 Hysteresis Loss

In type-II superconductors, the magnetic field exists in the form of magnetic flux vortices which are pinned in the defects of the material. The vortices thus do not move under the Lorentz force perpendicular to the transport current. That is why a hard type-II superconductor can carry a large constant transport current with zero resistance. However, the presence of changing currents or fields makes the magnetic flux lines move. A voltage is induced, which results in dissipation due to the normal regions associated with the vortices.

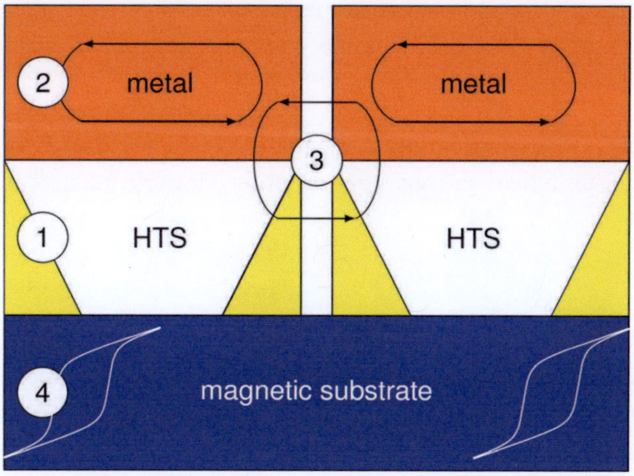

Figure 2.8: Schematic of losses in an ReBCO HTS tape. 1) Hysteresis loss in the superconducting material; 2) eddy current loss in the normal metallic components; 3) coupling loss between filaments; 4) ferromagnetic loss in the magnetic substrate if applicable [33].

Macroscopically, the magnetization loop of a type-II superconductor is hysteretic as described in Chapter 2.4.1. The area of the hysteresis loop (Fig. 2.5) is the dissipated energy during a magnetization cycle

$$Q_{\mathrm{hts,c}} = \int H \cdot dM = \int M \cdot dH. \tag{2.7}$$

The AC loss in the type-II superconductor based on the critical state model can be solved analytically for specific geometries, which will be presented in Chapter 3.1. Based on the critical state model which considers quasi-static conditions and firm flux pinning, the AC loss per cycle is independent on the rate of the variation of the transport current or the external field.

Extra loss can be caused due to flux creep or flux flow, which can be further considered using the *E-J* power law. Moreover, the magnetic field and temperature dependence of J_{c} will also impact the AC loss. These can be solved using numerical methods.

2.5.2 Eddy Current Loss

HTS tapes are based on alloy substrates and normally covered with metallic stabilizers. When an HTS tape is subject to a varying magnetic field or carrying a varying current, eddy currents will be generated in metallic parts and generate dissipation according to Faraday's law.

When a thin metal sheet of thickness d (on the top of in Fig. 2.4) is subject to a quasi-static sinusoid magnetic field, the energy dissipation per cycle per unit volume is [46]

$$Q_{\mathrm{metal,c}} = \frac{\pi^2 B_{\mathrm{m}}^2 d^2 f}{6\rho_{\mathrm{e}}}, \tag{2.8}$$

where B_{m} is the amplitude of the applied field, f is the frequency and ρ_{e} is the electrical resistivity of the metal. Unlike the hysteresis losses in the superconducting material under the critical state model assumptions, the energy dissipation per cycle in normal metal is frequency dependent.

To reduce the eddy current loss for applications such as transformers, thinner metal sheets, larger resistivity and slower field are preferred. The strands of cables are often twisted to reduce enclosed magnetic flux. On the other hand, eddy current loss can be used for induction heating of metals.

In HTS tapes or devices, the superconducting and metallic parts influence each other though electromagnetic fields. Moreover, the geometry is complex and the material properties are temperature dependent or not uniform. The eddy current loss can be calculated numerically together with the hysteresis loss in superconducting parts.

2.6 Trapped-Field Magnets

By virtue of strong flux pinning, HTS bulks or composite bulks (stacks of coated conductors) are able to provide strong magnetic fields, thus are promising to improve the performance of many applications. These applications can be divided into two groups according to the function of the HTS components:

1) HTS (composite) bulks act passively with respect to other permanent magnetic parts. Typical applications include maglev (magnetic levitation) vehicles [47], [48], [49], magnetic bearings [50], flywheel energy storage devices [51] and hysteresis or reluctance rotating machines [52].

2) HTS bulks or stacks of coated conductors work as quasi-permanent magnets after magnetization (as described in Chapter 2.4.1). Such magnets are known as trapped-field magnets (TFMs) and can be used in a variety of applications that use conventional permanent magnets such as rotating machines [52], nuclear magnetic resonance magnets [53], [54] and magnetic separators [55], [56].

This thesis focuses on the magnetization techniques of trapped-field magnets for practical applications. Much work has been reported for the magnetization of HTS bulk trapped-field magnets [2]. In recent years, HTS stacks of coated conductors have been attracting more and more interest. Compared to HTS bulks, stacks of HTS coated conductors have several advantages: there are more manufactures of HTS coated conductors and the stacks are relatively cheaper; the mechanical strength of stacks is better due to the metallic substrates; and the shapes of stacks are more flexible to adapt to different configurations of applications [49].

2.6.1 Applications

Trapped-field magnets can trap magnetic fields an order of magnitude higher than those produced by conventional permanent magnets, which are typically less than 2 T [57]; as a result, the replacement of conventional magnets with trapped-field magnets in rotating machines is attracting great interest. There are already designs and demonstrations of rotating machines with HTS bulk trapped-field magnets, some of which can be found in [52]. Since trapped-field magnets provide much higher magnetic flux, trapped-field magnets are promising for increasing the power or torque density and reducing the size or weight of the machine.

A conceptual drawing of a motor that uses long rectangular stacks of coated conductors as rotor poles was given in [58]. Instead of bulks that are usually present as cylinders, stacks of coated conductors are more flexible to suit for the configuration and can provide more magnetic flux by increasing the length of the rotor. The stator coils can also be used to magnetize the stators to realize *in situ* magnetization, which will be introduced in Chapter 2.6.2.2.

Trapped-field magnets are also promising to be used as nuclear magnetic resonance (NMR) magnet to realize a more compact design [53], [54]. Trapped-field magnets can provide a stationary background field after magnetization so that high current leads and liquid helium supply systems are no longer needed. Much work has been done to improve the spatial homogeneity and temporal stability of trapped magnetic fields [59], [60].

Trapped-field magnets can be used in magnetic separators for water purification and ore selection [55], [56]. A TFM can provide magnetic fields of a much larger gradient on the surface compared to a conventional solenoid. Since the forces acting on magnetic particles are the products of the magnetization of particles and the gradient of the magnetic field, the magnetic separator with trapped-field magnets can provide larger separation forces and thus be more effective [55].

2.6.2 Magnetization

Effective, cheap and stable magnetization is the key issue towards practical applications of trapped-field magnets. There are generally three techniques for magnetization of HTS (composite) bulks: field cooling (FC), zero field cooling (ZFC), pulsed field magnetization (PFM). Field cooling and zero field cooling are also known as static field magnetization methods relative to PFM. Another novel concept, "flux-pump", is also drawing some attention [61], [62], [63]. So far, PFM is the most promising method that can realize *in situ* magnetization of trapped-field magnets in rotating machines.

2.6.2.1 Static Field Magnetization

Static field magnetization means that the variation rate of the magnetic field during the magnetization process is so low that the magnetization is quasi-static and the temperature is kept constant. The most commonly employed static field magnetization techniques are field cooling (FC) and zero field cooling (ZFC) depending on whether the superconductor is cooled below its critical temperature in the presence of an external field or not.

Table 2.1: Representative trapped fields of trapped-field magnets obtained by field cooling or zero field cooling method

Trapped field	Temperature	Material	Geometry	Reference
17.6 T	26 K	GdBCO	2 bulks, $\phi = 25$ mm	[1]
17.24 T	29 K	YBCO	2 bulks, $\phi = 26.5$ mm	[64]
3.05 T	77 K	GdBCO	bulk, $\phi = 65$ mm	[65]
5.4 T	12 K	MgB_2	bulk, $\phi = 30$ mm	[66]
7.9 T	4.2 K	GdBCO	stack of 260-layer coated conductors, 12 mm square	[67]
7.34 T	4.2 K	YBCO	stack of 120-layer coated conductors, 12 mm square	[68]

ϕ: diameter

During zero field cooling and field cooling, the magnetic field and current distribution of a superconducting slab can be described by the critical state model, as shown in Fig. 2.4. A quasi-static external magnetic field of magnitude two (zero field cooling) or one (field cooling) times of the trapped fields has to be applied to realize full magnetization. Static field magnetization methods can magnetize a superconductor to its theoretical maximum magnetization when the current density everywhere inside a superconductor reaches J_c. Some representative trapped fields produced by static field magnetization are given in Tab. 2.1.

Static field cooling needs to use a superconducting magnet to supply a high field with a low changing rate. In [1], a magnetic field up to 18 T was applied when the sample was maintained at 100 K, and then the sample was cooled to 26 K. Finally, the external field was reduced to zero at a damping rate of 0.015 T/min. The operation took more than 20 hours. Moreover, it is impossible to refill the magnetization if the trapped field is degraded by disturbance in the environment. In a practical assembly, it is not realistic to include a superconducting magnet with its cryogenic attachments for magnetization of trapped-field magnets. A possible solution is to transfer a trapped-field magnet to the application after magnetization. Then the transportation and magnetization have to be done each time when the application is warmed up above the HTS operating temperature. Overall, the field cooling and zero field are not practical for applications.

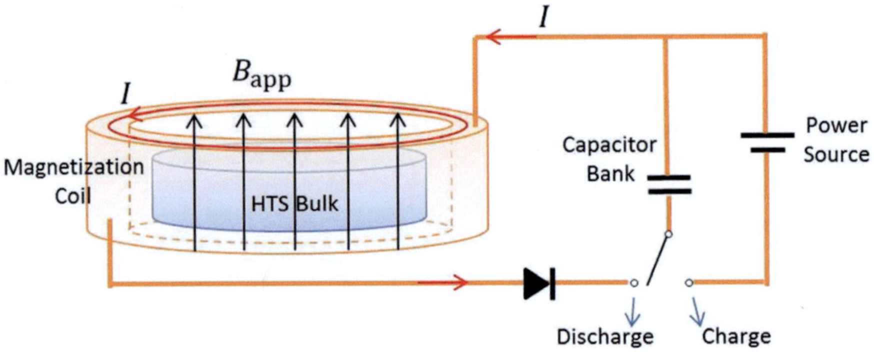

Figure 2.9: Schematic illustrations of a PFM apparatus to magnetize an HTS bulk immersed in liquid nitrogen [69].

2.6.2.2 Pulsed Field Magnetization

Pulsed field magnetization (PFM) is attracting attention as a promising solution for *in situ* magnetization for trapped-field magnets. Instead of a quasi-static magnetic field, a pulsed magnetic field is applied to magnetize the sample below its critical temperature.

An example of the set-up of PFM is shown in Fig. 2.9. The magnetization coil is made of copper. A capacitor bank is charged and then discharges a pulsed current to the magnetization coil. The pulsed current usually lasts for tens of milliseconds depending on the parameters of the circuit. For the critical damping situation (as shown in Fig. 2.10), the shape of the pulse can be represented by an exponential function [21]

$$B_{\mathrm{app}}(t) = B_{\mathrm{m}} \frac{t}{\tau} \mathrm{e}^{1-\frac{t}{\tau}}, \tag{2.9}$$

where τ is the characteristic time ($\tau = 12$ ms in Fig. 2.10).

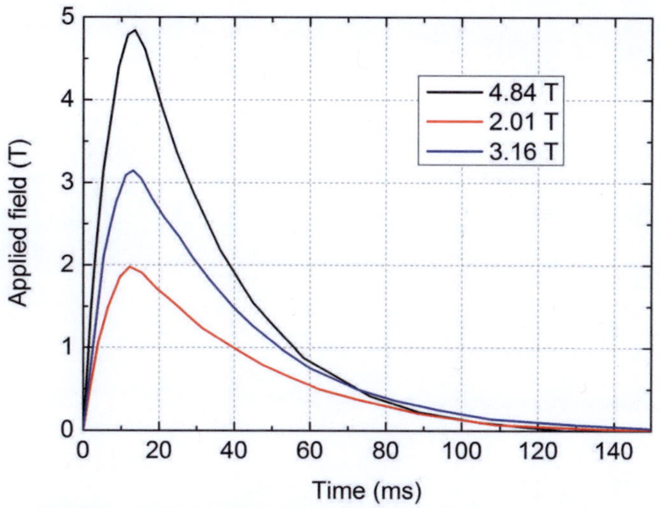

Figure 2.10: Typical applied pulsed fields with time (measured) [21].

One of the key issues of PFM is that the produced trapped field cannot reach the full potential of the trapped-field magnet as field cooling can produce, especially at low temperatures. The fast flux motions during PFM generate excessive heat and increase the temperature, as a result, the critical current density is decreased and the produced trapped field is substantially degraded [2]. The remarkable high trapped fields in Tab. 2.1 are all produced by field cooling or zero field cooling. For comparison, the world record trapped field produced by PFM is only 5.9 T for HTS bulks [72] and 2 T for stacks of coated conductors [73]. Effort has been devoted to understand the flux dynamics during PFM and optimize the process to achieve more trapped fields by PFM. The effort is devoted mainly into three aspects, the coil configurations, multi-pulse technique and composite bulk structures [2].

Vortex coils (as shown in Fig. 2.11) were found to show several advantages over common solenoids. It was shown that PFM by vortex coils could produce more trapped field in an HTS bulk than by a solenoid in experiments [74]. This was explained as the result of reduction of heat generation on the periphery of the bulk [75]. The cooling from the side of the cylinder could also be more efficient for the vortex coils arrangement because the thermal conductivity along the *ab*-plane of the bulk is larger than that of along the *c*-axis [76]. Moreover, the vortex coils can also be used as stators and the arrangement can be more suitable for a practical motor [58], [77].

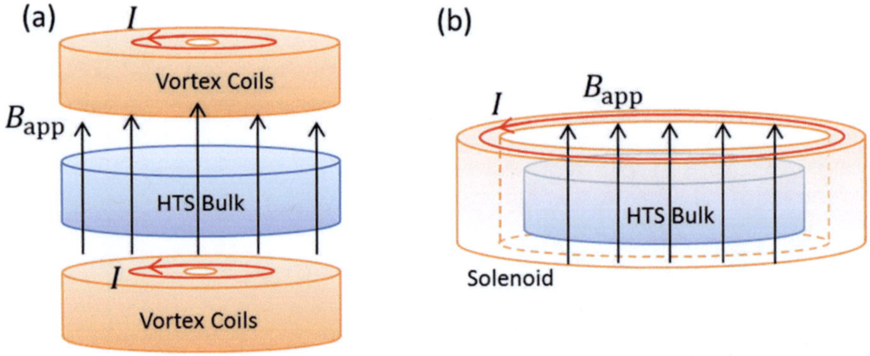

Figure 2.11: Illustration of the magnetization coils for PFM: (a) a pair of vortex coils; (b) a common solenoid [74].

The multi-pulse technique was proved to be effective for improving the trapped field of HTS bulks by PFM experimentally. The record trapped fields in [72] and [73] were acquired by using multi-pulse techniques. The multi-pulse techniques include successive pulsed-field application (SPA), iteratively magnetizing pulsed field method with reducing amplitude (IMRA), and multi-pulse technique with step-wise cooling (MPSC) [78], [79], [80], [81]. These techniques are illustrated in Fig. 2.12. For SPA and IMRA pulses of repeating or reducing amplitudes are applied at a fixed temperature; for MPSC, the sample is first magnetized with pulses of smaller amplitudes at a higher temperature, and then cooled to an even lower temperature and magnetized with pulses of larger amplitudes. The increase of the trapped field by successive pulses was understood as a result of reducing temperature rise due to the existence of currents produced by previous pulses. The mechanism of the improvement in the trapped field and flux by multi-pulse technique is not clear yet. The applied pulse sequences used in experiments are usually determined empirically. Little numerical simulation of multi-pulse technique has been reported and they have been carried out only for HTS bulks [82], [83].

In terms of composite bulk structures, HTS bulks with metallic composite structures to improve the thermal conductivity are investigated numerically and experimentally [84], [85], [86], [87]. The primary results demonstrate the benefits of composite structures of HTS bulks, which suggest that stacks of HTS coated conductors as composite bulks have advantages naturally in terms of improved mechanical strength, higher thermal conductivity and more homogeneous critical current density [2].

The aim of this thesis is to contribute to understanding the flux dynamics in trapped-field magnets during PFM and furthermore to propose possible strategies to improve trapped fields produced by PFM. The research is manly based on numerical simulation with electromagnetic-thermal coupled models and experiments are carried out for validation. The main contributions of the thesis are simulations of stacks of coated conductors magnetized by PFM, which are rarely reported in the literature. Since the research in the literature on HTS bulks are more extensive and the geometry and structures are relatively simple, the HTS bulks are first simulated to understand the fundaments of the flux dynamics during PFM.

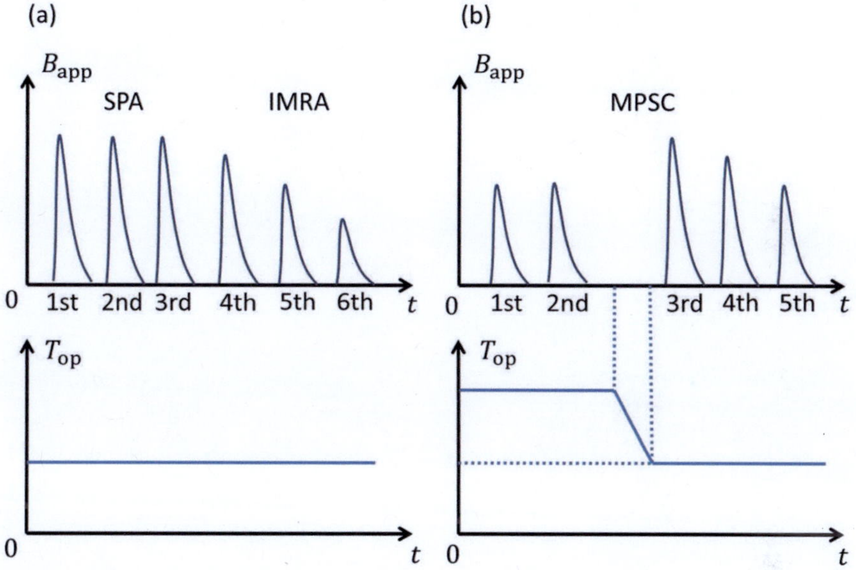

Figure 2.12: Schematic diagrams of the applied field and corresponding cooling temperatures of PFM with multi-pulse technique: (a) successive pulsed-field application (SPA) and iteratively magnetizing pulsed field method with reducing amplitude (IMRA); (b) multi-pulse technique with step-wise cooling (MPSC).

3 Modelling of Superconductors

Numerical modelling is a powerful tool in terms of both understanding mechanisms behind physical phenomenon and predicting performances of applications, especially for complex systems where analytical solutions are not applicable. The properties of superconductors strongly depend on the temperature and orientation and magnitude of the magnetic field; in turn, the temperature varies temporally and spatially due to generated heat or cooling. Numerical modeling considering the coupling of electromagnetism and temperature is therefore necessary for the simulation of pulsed field magnetization when temperature changes greatly. As the basis of this thesis, this chapter describes the fundamentals of numerical modelling of type-II superconductors when electromagnetism and temperature are coupled.

3.1 Analytical Models

Based on the critical state model [38], [39], there are analytical equations to describe electromagnetism of type-II superconductors of simple geometries and constant critical current density. Specifically, Brandt's model (Chapter 3.1.1) [88], [89], [90], [91], [92] (and the equivalent derivation by Zeldov *et al* [93]) describes the magnetic field and current distribution in and around a superconductor of infinite length when it is exposed to a varying external applied field. On the other hand, the Norris's model (Chapter 3.1.2) considers the situation when a superconductor of infinite length is carrying alternating currents.

3.1.1 Brandt's Model

This thesis mostly focuses on the magnetization of HTS trapped-field magnets, which is relevant to Brandt's model for the most ideal cases describing field cooling or zero field cooling. The Brandt's model can account for arbitrary cross section and the *E-J* power law, including the critical state limits when the exponent index n approaches infinity. When a superconductor is exposed to a varying external magnetic field, currents are induced in the superconductor opposing the change of the magnetic flux in it. The Maxwell's equations have to be fulfilled during the process. With Green function method [94], the current penetration and field evolution can be calculated analytically or numerically [88], [89]. For the critical state limits, the current density in the superconductor can only be zero or the critical current density and simple analytical solutions are available in some circumstances.

Although the idea is straightforward, the mathematical derivations are not trivial. Only superconductors of certain geometries can be solved analytically. Many papers are published solving the critical states in superconductors of different geometries theoretically, represented by [88], [89], [90], [91], [92]. The mathematical derivations will not be shown here. A concise summary of field and current distribution of an infinitely long strip superconductor can be found in [95].

The penetration field H_p is the minimum external field that can fully magnetize the superconductor by field cooling according to the critical state model. Besides, H_p is also the magnetic field in the center of the superconductor when it is fully magnetized, so H_p can also represent the capacity of the superconductor in trapping magnetic field. This value is therefore of special interest as a theoretical benchmark for the numerical modelling of trapped-field magnets.

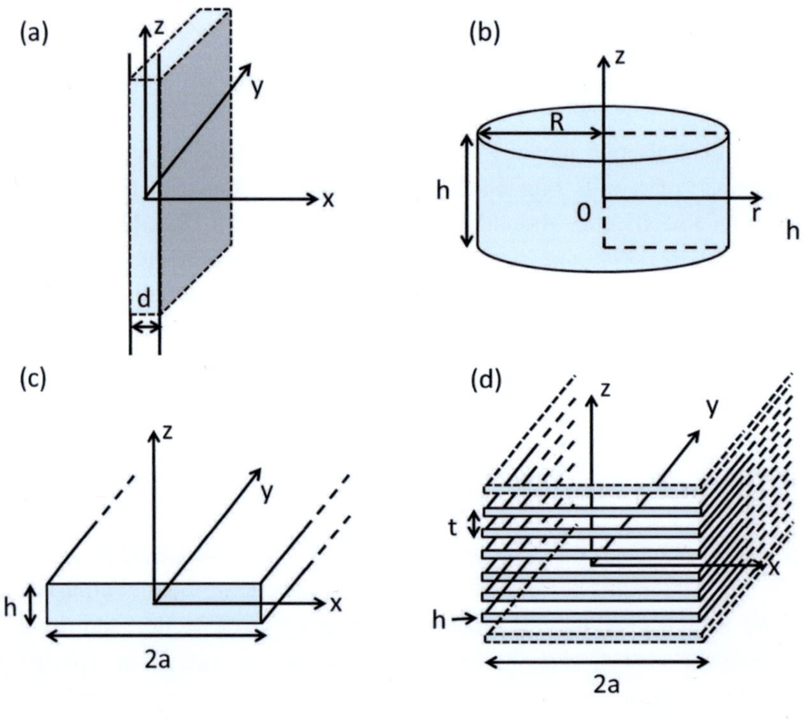

Figure 3.1: Geometries that can be analytically solved with Brandt's model with constant J_c for the penetration field H_p and the magnetization loss per volume per cycle Q_c as listed in Tab. 3.1. The external field is along the z direction.

The penetration field H_p and the magnetization loss per volume per cycle Q_c for superconductors of particular geometries shown in Fig. 3.1 are listed in Tab. 3.1. The AC external magnetic field is along the z-axis direction and the amplitude is H_m. Among the different geometries, the cylinder is the typical geometry of an HTS bulk and the strip is that of a ReBCO HTS tape. The magnetization loss of a strip is characterized by a fourth-power dependence on H_m for low applied field, and by a first-power dependence for high field. The stack of strips behaves similar to a slab when the distance of strips t is much smaller than the strip width $2a$.

3.1.2 Norris's Model

When the superconductor is carrying alternating current, the magnetic field and current distribution is described by the Norris's model [96] and the hysteresis loss can be calculated analytically. The Norris's model is applicable for applications in which superconductors are used for carrying transport current such as cables or coils.

The model also includes the critical state model assumption and the critical current density is magnetic field independent. When the total current in a superconductor is changed, the current distribution on the periphery starts to change first. The magnetic field and current distribution always tends to hinder the change of magnetic flux inside. The current density in the superconductor is either zero or the critical current density. The current distribution is solved by properly filling currents into the cross-section with conformal mapping techniques. The hysteresis loss per cycle per unit length in a wire of elliptic cross-section or in a strip (as shown in Fig. 3.2) when carrying an AC current is

$$Q_{\text{strip}} = \frac{I_c^2\mu_0}{\pi}\left\{\left(1-\frac{I}{I_c}\right)\ln\left(1-\frac{I}{I_c}\right)+\left(2-\frac{I}{I_c}\right)\frac{I}{2I_c}\right\}, \quad \textbf{(3.1a)}$$

$$Q_{\text{elliptic}} = \frac{I_c^2\mu_0}{\pi}\left\{\left(1-\frac{I}{I_c}\right)\ln\left(1-\frac{I}{I_c}\right)+\left(1+\frac{I}{I_c}\right)\ln\left(1+\frac{I}{I_c}\right)-\left(\frac{I}{I_c}\right)^2\right\}, \quad \textbf{(3.1b)}$$

where I_c is the critical current of the wire. For low current, the loss of strips (Eq. 3.1a) has a fourth-power dependence on current for low current, while that of elliptic cross-section wires (Eq. 3.1b) has a third-power dependence.

Table 3.1: The penetration field H_{p} and the magnetization loss per volume per cycle Q_{c} for different geometries as illustrated in Fig. 3.1 derived by the critical state model with constant J_{c}. The amplitude of the AC external field is H_{m}.

Geometry	H_{p}	Q_{c}	Reference
(a) Slab	$\frac{J_{\mathrm{c}}d}{2}$	$\frac{2\mu_0 H_{\mathrm{m}}^3}{3H_{\mathrm{p}}}$, if $H_{\mathrm{m}} < H_{\mathrm{p}}$ $\frac{\mu_0 H_{\mathrm{p}}}{3}(6H_{\mathrm{m}} - 4H_{\mathrm{p}})$, if $H_{\mathrm{m}} \geq H_{\mathrm{p}}$	[40], [33]
(b) Cylinder	$\frac{J_{\mathrm{c}}h}{2}\ln\left(\frac{2R}{h} + \sqrt{\left(1 + \frac{4R^2}{h^2}\right)}\right)$	Numerical	[40], [90]
(b) Disk h ≪ 2R	$\frac{J_{\mathrm{c}}h}{2}$ (*)	Numerical	[40], [97]
(c) Tape	$\frac{J_{\mathrm{c}}h}{2\pi}(\frac{4a}{h}\arctan\left(\frac{h}{2a}\right) + \ln\left(1 + \frac{4a^2}{h^2}\right))$	Numerical	[40], [98]
(c) Strip h ≪ 2a	$\frac{J_{\mathrm{c}}h}{\pi}$ (*)	$2a\mu_0 J_{\mathrm{c}} H_{\mathrm{m}}\{\frac{2H_{\mathrm{p}}}{H_{\mathrm{m}}}\ln\left(\cosh\left(\frac{H_{\mathrm{m}}}{H_{\mathrm{p}}}\right)\right) - \tanh\left(\frac{H_{\mathrm{m}}}{H_{\mathrm{p}}}\right)\}$	[40], [33], [99]
(d) Stack of strips $h \ll 2a$	$\frac{J_{\mathrm{c}}h}{\pi}$ (*)	$\frac{2\mu_0 t^2 H_{\mathrm{m}}^2}{\pi a h}\int_0^1 (1-2s)\ln(1 + \frac{\sinh^2(\frac{\pi a}{d})}{\cosh\left(\frac{\pi H_{\mathrm{m}} s}{J_{\mathrm{c}} h}\right)})ds$	[40], [100]

(*) Indicating that the penetration field is mathematically infinite, so the field for significant penetration is given.

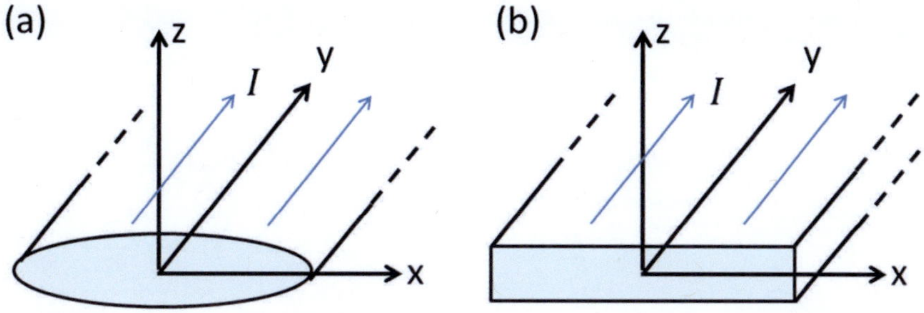

Figure 3.2: Schematics for infinitely long HTS tapes carrying transport currents along the y direction: (a) an HTS wire of elliptical cross-section and (b) a strip.

3.2 Overview of Numerical Models

Numerical models are capable of solving problems with increasing complexity, so that arbitrary geometries can be simulated taking into account the temperature and magnetic field dependences of the critical current density. Moreover, numerical models make it possible to solve coupled electromagnetic and thermal problems.

A numerical model includes representing a physical behavior with simplified assumptions, generating equations that approximate the model, expressing the model in a discrete form, and solving the equations [32], [101].

This thesis focuses on investigating the dynamic process of magnetization while electromagnetics and temperature are coupled. The magnetic field and current distributions are calculated to determine the energy dissipation in superconductors. In this sense, the model is built with a realistic or relatively reduced geometry, with the assumption of the critical state model or the *E-J* power law. The magnetic field dependence and position dependence of the critical current can be taken into account to provide more accurate simulation. Other models are available to consider large-scale applications when detailed current and field distributions are not required, e.g. circuit models [102], [103], [104].

The model is expressed in mathematical forms (namely formulations) to be solved numerically. They can be categorized according to the type of equations solved: partial differential equations (PDEs), integral equations (IEs) or others, which can be found in several exhaustive reviews [32], [33], [101], [105], [106].

Table 3.2: The most common formulations for modelling superconductors[105], [106]

Formulation	Definition of state variables	PDE solved for
$\boldsymbol{A}-\varphi$	$\boldsymbol{B}=\nabla\times\boldsymbol{A}$ $\boldsymbol{E}=-\frac{d\boldsymbol{A}}{dt}-\nabla\varphi$ $\sigma=\sigma(E)$	$\nabla\times\left(\frac{1}{\mu}\nabla\times\boldsymbol{A}\right)=-\sigma(\frac{d\boldsymbol{A}}{dt}+\nabla\varphi)$ $\nabla^2\varphi=0$
$\boldsymbol{T}-\Omega$	$\boldsymbol{J}=\nabla\times\boldsymbol{T}$ $\boldsymbol{H}=\boldsymbol{T}-\nabla\Omega$ $\rho=\rho(J)$	$\nabla\times(\rho\nabla\times\boldsymbol{T})=-\mu(\frac{d\boldsymbol{T}}{dt}-\nabla\frac{d\Omega}{dt})$ $\nabla^2\Omega=0$
$\boldsymbol{H}$	$\boldsymbol{J}=\nabla\times\boldsymbol{H}$ $\rho=\rho(\boldsymbol{J})$	$\nabla\times(\rho\nabla\times\boldsymbol{H})=-\mu\frac{d\boldsymbol{H}}{dt}$

The finite element method (FEM) is the most used and developed method to solve PDEs, because of its generality to solve arbitrary shapes in any dimensions and wide availability of commercial codes. The FEM is characterized by the fact that it does not solve directly the PDEs, but rather a modified version of them, called "weak form", which consists of an integral over the whole domain of the product of the original PDEs multiplied by so-called "weighting functions". The latter are used to weight the error over the domain, which opens all sorts of possibilities in term of numerical resolution. When compared with other methods, for instance the finite difference method, which solves exactly a discretized version of the original PDEs (called "strong form") at specific points in space without controlling the error between these points, the FEM allows spreading the approximation error in a controlled manner over the whole domain. In addition, since the FEM supports unstructured meshes, it makes it very general to solve problems with complicated geometries, which is not the case with most strong form methods. The inherent smoothness of finite elements also handles discontinuities and nonlinear material properties much better than many other numerical methods. When modeling superconductors, the associated FEM problem is primarily an electromagnetic one, which dictates commands for the use of formulations such as $\boldsymbol{A}-\varphi$, $\boldsymbol{T}-\Omega$ and $\boldsymbol{H}$-formulations as listed in Tab. 3.2. Other formulations, including E or J formulations are also available [107], [108].

3.3 *H*-Formulation of Maxwell's Equations

The ***H***-formulation uses the magnetic field ***H*** as the state variable for solving Maxwell's equations. Since it was proposed [109], [110], the ***H***-formulation has been mostly used for calculating the hysteresis losses of superconductors. The use of linear edge elements [111] improves the ***H***-formulation further in stability and speed.

The FEM simulation in this thesis is carried out with the commercial software COMSOL Multiphysics [112], which provides easy use of edge elements and other necessary tools. In the following, the implementation of the equations used in this thesis is introduced. Other aspects of the model will be discussed in later chapters including geometry, mesh, integral constraints, and solvers.

Starting from the Maxwell's equations with no magnetic materials

$$\nabla \times \boldsymbol{E} = -\mu \frac{\partial \boldsymbol{H}}{\partial t}, \tag{3.2}$$

$$\nabla \times \mu \boldsymbol{H} = -\mu(\boldsymbol{J} + \epsilon_0 \frac{\partial \boldsymbol{E}}{\partial t}), \tag{3.3}$$

and ignoring the term $\partial \boldsymbol{E}/\partial t$ for low-frequency problems, one can easily come to the equation with state variable $\boldsymbol{H}$

$$\nabla \times (\rho \nabla \times \boldsymbol{H}) = -\mu \frac{\partial \boldsymbol{H}}{\partial t}, \tag{3.4}$$

where the resistivity of the superconductor is assumed to follow the power law

$$\rho = \frac{E_c}{J_c}\left(\left|\frac{\boldsymbol{J}}{J_c}\right|\right)^{n-1}, \tag{3.5}$$

and $\boldsymbol{J}$ and $\boldsymbol{H}$ are related by

$$\boldsymbol{J} = \nabla \times \boldsymbol{H}. \tag{3.6}$$

3.3.1 2D Planar Model

Objects of infinite length, for example, long tapes or stacks as shown in Fig. 3.1(c) and (d), can be considered with 2D planar models when the physics is the same along a certain direction. As shown in Fig. 3.3(a), the Cartesian coordinate is used. The magnetic fields only have x- and y- components and the current density only have z-component

$$\mu \frac{\partial H_x}{\partial t} + \frac{\partial E_z}{\partial y} = 0, \tag{3.7a}$$

$$\mu \frac{\partial H_y}{\partial t} - \frac{\partial E_z}{\partial x} = 0, \tag{3.7b}$$

$$E_z = \rho J_z = \rho\left(\frac{\partial H_y}{\partial x} - \frac{\partial H_x}{\partial y}\right). \quad \textbf{(3.8)}$$

In COMSOL Multiphysics [112], the General Form PDE (g) module is used. The equations take the forms:

$$e_a \frac{\partial^2 u}{\partial t^2} + d_a \frac{\partial u}{\partial t} + \nabla \cdot \Gamma = f\,, \quad \textbf{(3.9a)}$$

$$u = \left[H_x, H_y\right]^T, \quad \textbf{(3.9b)}$$

$$\nabla = \left[\frac{\partial}{\partial x}, \frac{\partial}{\partial y}\right], \quad \textbf{(3.9c)}$$

where e_a, d_a and f are coefficient matrixes. In order to adapt Eq. 3.7 into Eq. 3.9, one needs to fill the terms in the user interface as

$$\Gamma = \begin{pmatrix} 0 & E_z \\ -E_z & 0 \end{pmatrix}, \quad \textbf{(3.10a)}$$

$$d_a = \begin{pmatrix} \mu & 0 \\ 0 & \mu \end{pmatrix}, \quad \textbf{(3.10b)}$$

where f and e_a are all zero. In the Definitions section, E_z is defined for different domains with different resistivity ρ according to the corresponding materials.

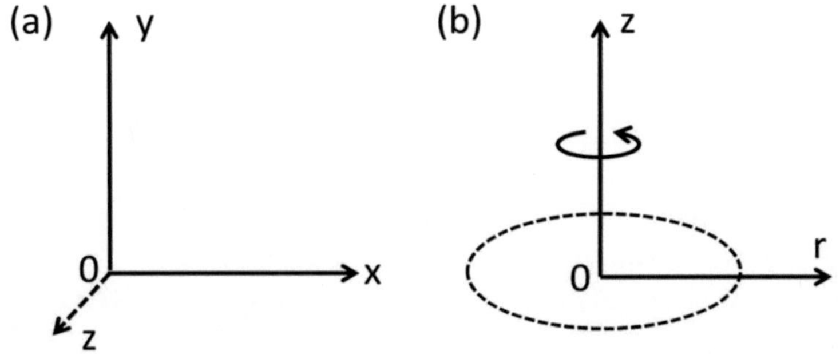

Figure 3.3: Two-dimensional coordinate systems: (a) a 2D Cartesian coordinate system; (b) a cylindrical coordinate system.

3.3.2 2D Axisymmetric Model

Axisymmetric objects, for example, a cylindrical bulk as shown in Fig. 3.1(b), can be considered with 2D axisymmetric models. In a cylindrical coordinate as shown in Fig. 3.3(b)

$$\mu_0 \frac{\partial H_r}{\partial t} - \frac{\partial E_\phi}{\partial z} = 0 , \tag{3.11a}$$

$$\mu_0 \frac{\partial H_z}{\partial t} + \frac{1}{r}\frac{\partial (rE_\phi)}{\partial r} = \mu_0 \frac{\partial H_z}{\partial t} + \frac{\partial E_\phi}{\partial r} + \frac{E_\phi}{r} = 0 , \tag{3.11b}$$

$$E_\phi = \rho J_\phi = \rho \left(\frac{\partial H_r}{\partial z} - \frac{\partial H_z}{\partial r} \right). \tag{3.12}$$

To implement these equations in COMSOL, several approaches are available.

(A) General Form PDE (g) module is used. The equations take the forms

$$e_a \frac{\partial^2 u}{\partial t^2} + d_a \frac{\partial u}{\partial t} + \nabla \cdot \Gamma = f , \tag{3.13a}$$

$$u = [H_r, H_z]^T , \tag{3.13b}$$

$$\nabla = \left[\frac{\partial}{\partial r}, \frac{\partial}{\partial z} \right], \tag{3.13c}$$

where e_a, d_a and f are coefficient matrixes. In order to adapt Eq. 3.11 into Eq. 3.13, the blanks in the user interface should be filled as

$$\Gamma = \begin{pmatrix} 0 & -E_\phi \\ E_\phi & 0 \end{pmatrix}, \tag{3.14a}$$

$$d_a = \begin{pmatrix} \mu & 0 \\ 0 & \mu \end{pmatrix}, \tag{3.14b}$$

$$f = \begin{pmatrix} 0 \\ -\frac{E_\phi(J)}{r} \end{pmatrix}, \tag{3.14c}$$

while e_a is all zero. In the Definitions section, E_ϕ for different domains should be defined with different resistivity ρ according to their material types. The problem with this approach is that when r approaches zero, the source term f goes to infinite. This implementation method is not used because it produces errors in some situations.

(B) Another way to write the equations in General Form PDE (g) module is to take rE_ϕ as a whole to rewrite the equation as the left of Eq. 3.11b [113]. The source term f can be set to zero. One can fill in the blanks in the user interface as

$$\Gamma = \begin{pmatrix} 0 & -rE_\phi \\ rE_\phi & 0 \end{pmatrix}, \tag{3.15a}$$

$$d_a = \begin{pmatrix} r\mu & 0 \\ 0 & r\mu \end{pmatrix}, \tag{3.15b}$$

while f and e_a are all zero.

(C) In order to benefit from the AC/DC module in COMSOL, the ***H***-formulation PDEs can be correlated to the ***A***-formulation PDEs and solved with the Magnetic Fields (mf) module [114]. This approach is named as "hijack" model in the thesis. As shown in Tab. 3.3, the equations of ***H***-formulation and ***A***-formulation have the same format. By relating the variables in the ***H***-formulation with those in the ***A***-formulation, the PDEs of ***H***-formulation can be solved taking advantage of the Magnetic Fields (mf) module as a trick.

In order to implement the model in COMSOL, in the Materials section, the variable values σ and $1/\mu$ are defined according to the real values μ and ρ in the ***H***-formulation. In the boundary conditions, ***A*** on the boundary is defined according to the real values of ***H*** in the ***H***-formulation.

It is also possible to use the Magnetic Field Formulation (mfh) module to implement the ***H***-formulation directly. To take advantage of the AC/DC module in COMSOL and also keep track of the tradition of the research group, the 2D axisymmetric simulations in this thesis are carried out with approach (C) as described above.

Table 3.3: Correspondence of *H*-formulation and *A*-formulation to produce the "hijack" model

	H-formulation	***A***-formulation
Equations	$\nabla \times (\rho \nabla \times \boldsymbol{H}) = -\mu \frac{\partial \boldsymbol{H}}{\partial t}$	$\nabla \times \left(\frac{1}{\mu} \nabla \times \boldsymbol{A}\right) = -\sigma \frac{\partial \boldsymbol{A}}{\partial t}$
Variables	$\boldsymbol{H}$	$\boldsymbol{A}$
	μ	σ
	ρ	$1/\mu$
	$\boldsymbol{J}$	$\boldsymbol{B}$
	$\boldsymbol{E}$	$\boldsymbol{H}$

3.3.3 3D Model

In a 3D Cartesian coordinate system

$$\mu\frac{\partial H_x}{\partial t}+\frac{\partial E_z}{\partial y}-\frac{\partial E_y}{\partial z}=0\,, \tag{3.16a}$$

$$\mu\frac{\partial H_y}{\partial t}+\frac{\partial E_x}{\partial z}-\frac{\partial E_z}{\partial x}=0\,, \tag{3.16b}$$

$$\mu\frac{\partial H_z}{\partial t}+\frac{\partial E_y}{\partial x}-\frac{\partial E_x}{\partial y}=0\,, \tag{3.16c}$$

$$E_x=\rho J_x=\rho\left(\frac{\partial H_z}{\partial y}-\frac{\partial H_y}{\partial z}\right), \tag{3.17a}$$

$$E_y=\rho J_y=\rho\left(\frac{\partial H_x}{\partial z}-\frac{\partial H_z}{\partial x}\right), \tag{3.17b}$$

$$E_z=\rho J_z=\rho\left(\frac{\partial H_y}{\partial x}-\frac{\partial H_x}{\partial y}\right). \tag{3.17c}$$

The General Form PDE (g) module is used. The equations take the forms

$$e_a\frac{\partial^2 u}{\partial t^2}+d_a\frac{\partial u}{\partial t}+\nabla\cdot\Gamma=f\,, \tag{3.18a}$$

$$u=\left[H_x,H_y,H_z\right]^T, \tag{3.18b}$$

$$\nabla=\left[\frac{\partial}{\partial x},\frac{\partial}{\partial y},\frac{\partial}{\partial z}\right]. \tag{3.18c}$$

To adapt Eq. 3.16 into Eq. 3.18, fill in the terms in the user interface as

$$\Gamma=\begin{pmatrix}0 & E_z & -E_y\\ -E_z & 0 & E_x\\ E_y & -E_x & 0\end{pmatrix}, \tag{3.19a}$$

$$d_a=\begin{pmatrix}\mu & & \\ & \mu & \\ & & \mu\end{pmatrix}, \tag{3.19b}$$

while f and e_a are all zero. In the Definitions section, E_z should be defined for different domains with different resistivity ρ according to the corresponding materials.

3.3.4 Experimental Validation of 2D Planar Model

In order to validate the ***H***-formulation model in terms of predicting the AC loss, a single HTS tape is measured and compared with numerical simulation.

The sample is a 2G (GdBCO) coated conductor tape produced by SuNAM [115] (SLBK04150). It is 4 mm wide and overall 0.35 mm thick with Kapton tape insulation. The HTS is based on 100-µm-thick Hastelloy and laminated with brass (60 µm) on both sides. The self-field critical current of the sample is 172 A at 77 K.

The transport AC loss of the single tape at 77 K (immersed in liquid nitrogen) was measured with two different experimental set-ups independently and compared with the Norris curve (Chapter 3.1.2).

(A) Measurements at Shanghai Jiaotong University (SJTU) [116], [117]. The sample was supplied with a short AC current of only several cycles produced by a program-control current source. The current measured with a current clamp and the voltage between two voltage-taps are filtered, amplified and captured by an oscilloscope. The AC loss per cycle is calculated by direct integration of the product of voltage and current in a cycle.

(B) Measurements at Karlsruhe Institute of Technology (KIT) [114]. The sample was supplied with a continuous current. The current taken by a Rogowski coil and the voltage are measured with a lock-in amplifier to show the resistive loss power.

As shown in Fig. 3.4, the AC loss per cycle is independent on frequency, which reveals that the measured loss is of hysteretic nature. The results by the two methods reproduce each other at high currents where they both fit the Norris curve. At low currents, the measured values are substantially higher than the Norris ones. This difference is suspected to be related to non-uniform critical current density distribution along the width of the tape [118], [119], [120], [121]. The critical current density is lower on the edges of the tape due to the manufacturing process. To investigate the influence of this factor, a simplified expression is used [33]

$$J_c(x) = J_{c0} \times \begin{cases} 1, & \text{if } x \leq \dfrac{h_x w}{2} \\ 1 - \dfrac{(1-h_y)\left(2\left|\frac{x}{w}\right| - h_x\right)}{1-h_x}, & \text{if } x \leq \dfrac{h_x w}{2} \end{cases} \tag{3.20}$$

where w is the width of the tape. The profile of Eq. 3.20 is shown in Fig. 3.5. To make a fair comparison between different critical current density distributions, the integration of critical current density in the cross-section of the tape equals the self-field critical current I_{c0}

$$J_{c0} = \frac{2I_{c0}}{wd(1 + h_x + h_y - h_x h_y)}. \tag{3.21}$$

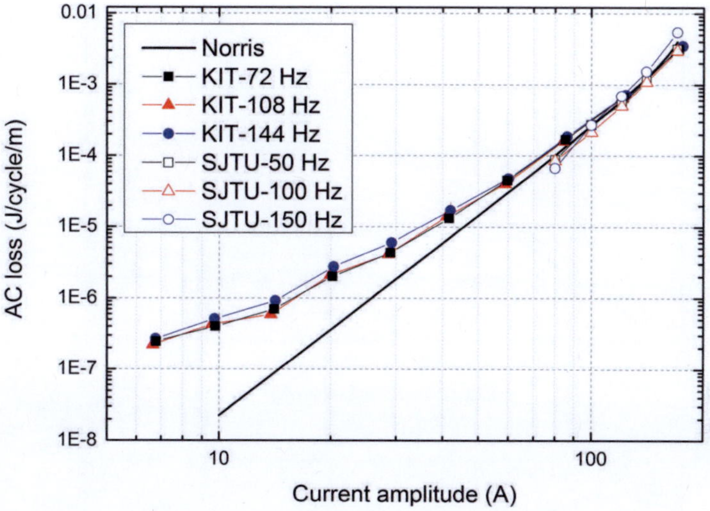

Figure 3.4: Transport AC loss of the single tape measured at KIT and SJTU.

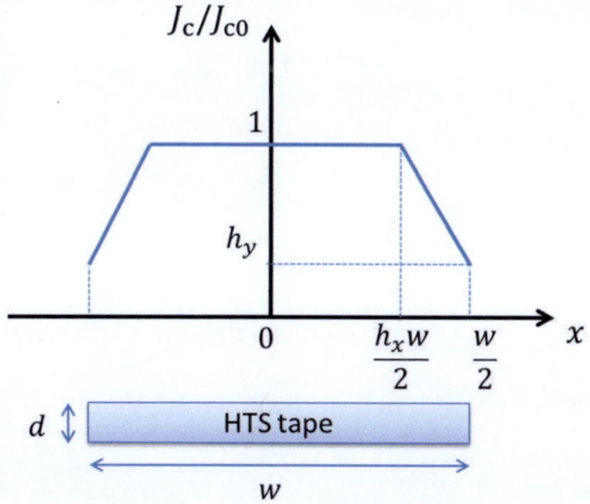

Figure 3.5: The simplified model for describing the non-uniform critical current density distribution along the width (x axis) of an HTS tape in Eq. 3.20.

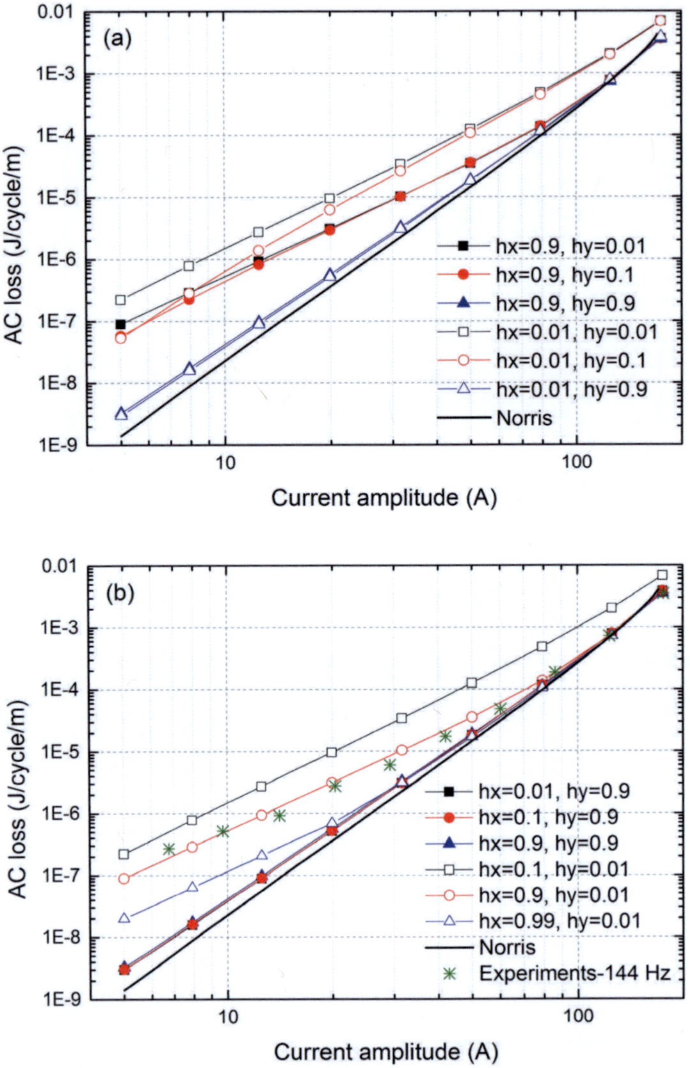

Figure 3.6: Calculated transport AC loss per cycle with the current amplitude for the same critical current (172 A). Different combinations of h_x and h_y in Eq. 3.20 are compared. For clarity, (a) compares different h_y when h_x is fixed and (b) compares different h_x when h_y is fixed.

For the same I_{c0}, the AC loss per cycle with the amplitude of applied current is calculated with the $\boldsymbol{H}$-formulation considering different combinations of h_x and h_y as shown in

Fig. 3.6. For clear visualization, Figure 3.6(a) compares different h_y when h_x is fixed and Figure 3.6(b) compares different h_x when h_y is fixed.

In both figures, the loss tends to converge to the Norris value at high currents. Figure 3.6(a) shows that when h_x is fixed, h_y influences the slope of the curve. In Figure 3.6(b), the curves show an increasing slope from low currents to high currents. And when h_y is fixed, h_x determines the changing point of the slope. Comparing with the experimental data, the calculated curve fits experiments when $h_x = 0.9$ and $h_y = 0.01$. This set of parameters is close to the experimental measurements in [120].

To confirm the influence of $J_c(x)$ on the AC loss, the magnetization loss of the same sample is measured and compared with calculation. The measurement was done at KIT using the calibration-free method [122]. The noise level is high at low applied fields. Generally, the calculation with $J_c(x)$ gives a better evaluation than the Brandt equation. At low applied fields, the magnetization loss is higher than the Brandt value because the low critical current region on the edges of the tape produces more losses when the magnetic fields penetrate [123]. This result confirms that the non-uniformity of critical current density along the width of the tape influences the AC loss of HTS tapes. In most power applications, the HTS tapes carry relatively high currents so the influence of non-uniformity is not a major concern. The non-uniformity should be considered in low current region to avoid underestimation of the AC loss.

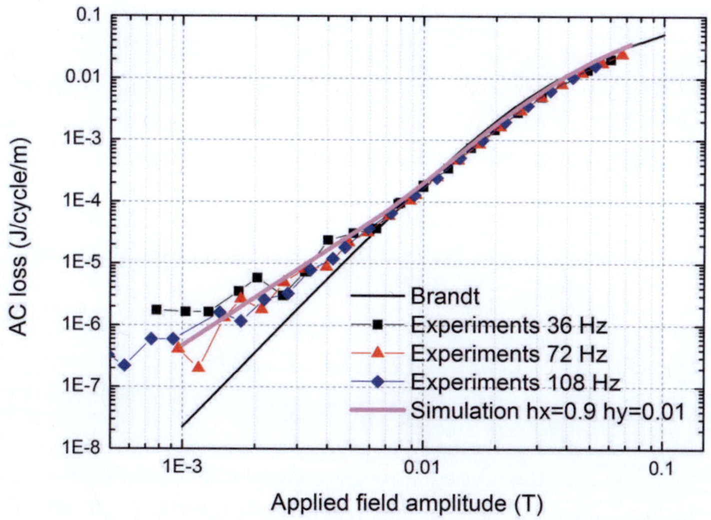

Figure 3.7: Measured magnetization AC loss and the calculated loss when $h_x = 0.9$ and $h_y = 0.01$.

3.4 Heat Transfer

When the critical current density is reached or exceeded, large energy dissipation is generated and the temperature increases; in turn, temperature greatly influences the critical current of superconductors and other properties of materials. As the superconducting parts are placed in a working environment with active cooling, the temperature is dynamically changed or stabilized. It is thus necessary to consider the heat transfer process.

In a superconductor, the temperature variation obeys the heat transfer equation [124],

$$\rho_m c_p \frac{\partial T}{\partial t} - \nabla \cdot (-\lambda \nabla T) = Q\,, \tag{3.22}$$

where ρ_m is the mass density, $c_p$is the heat capacity, λ is the thermal conductivity, and Q is the power density of the heat source, given by the Joule dissipation $\boldsymbol{E} \cdot \boldsymbol{J}$.

Two kinds of typical boundary conditions are used in this thesis, the Dirichlet boundary condition (Eq. 3.23a) and the Neumann boundary condition (Eq. 3.23b).

$$T|_{\text{boundary}} = T_0\,, \tag{3.23a}$$

$$\lambda \frac{\partial T}{\partial n}\bigg|_{\text{boundary}} = q_0. \tag{3.23b}$$

The Dirichlet boundary condition represents that the superconductor is attached directly or indirectly to a cold head of a given temperature. The Neumann boundary condition can be used when the cooling power per unit area q_0 (W/m^2) is known.

The solution of the heat transfer equation is commonly used to solve a variety of engineering problems. The Heat Transfer in Solid module in COMSOL is used directly.

3.5 Coupling of Heat Transfer and Electromagnetic Equations

The electromagnetic and heat transfer equations are coupled bi-directionally, as shown in Fig. 3.8. When the critical current density J_c is reached or exceeded in the superconductor, a large electric field $\boldsymbol{E}$ is generated according to the power law and hysteresis loss will be generated in the superconductor. The metallic components contained in HTS tapes have finite electrical resistivity ρ_e and generate eddy current loss when currents flow. The generated heat distributes and changes the temperature following the heat transfer equation. The temperature influences J_c and ρ_e and other material properties including the thermal conductivity λ and heat capacity c_p.

The coupling of equations is implemented by directly defining a heat source $\boldsymbol{E} \cdot \boldsymbol{J}$ in the Heat Transfer in Solids module. The temperature-dependent J_c, ρ_e, λ and c_p are defined with analytical equations or direct data interpolation.

It is worth pointing out that the calculation domains of the electromagnetic equations and thermal equations are not necessarily the same. The electromagnetic equations usually have a larger calculation domain than the thermal equations. The magnetic field close to a superconductor is usually not known before the calculation is done because the superconductor itself influences the excitation source. One has to create a large enough domain to impose the magnetic field as zero or an applied field on the outmost boundary. The temperature, however, can be imposed close to the superconductor (as described in Chapter 3.4), because the boundary condition is usually a Dirichlet boundary condition (the superconductor is attached to a given temperature) or a Neumann condition (the cooling power density is known). By reducing the domain of thermal equations to necessary parts (usually the superconductors only), the number of degrees of freedom is reduced and calculation time is saved.

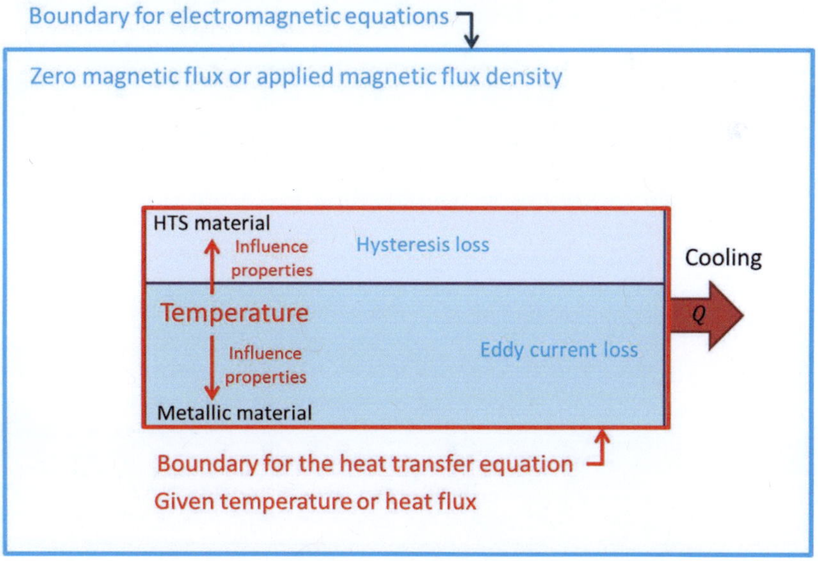

Figure 3.8: The coupling diagram of electromagnetic and thermal equations.

The electromagnetic part of the problem (***H***-formulation) is discretized with linear edge elements [111]. It requires a relatively finer mesh than that when using quadratic discretization. The heat transfer equations, as default, use quadratic discretization. However, when the electromagnetic and thermal equations are coupled, non-convergence appears. The temperature reaches a singularity at some point and the position of the error spot is mesh-dependent. This problem is solved by changing the discretization of the thermal equation from quadratic elements to linear elements. The stability is found to be greatly improved by this modification. The reason for this is not clear yet. One possibility is that using different orders of elements in two coupled physics may cause numerical mismatch. With a rather fine mesh (required by the ***H***-formulation), the use of linear instead of quadratic elements does not result in numerical inaccuracy. The simulation in this thesis is carried out using linear elements in both electromagnetic and thermal physics.

4 Magnetization of HTS Bulks

In this chapter, an electromagnetic-thermal coupled model is developed to describe HTS bulks magnetized by pulsed field magnetization (PFM). As a primary stage of the thesis, the initial motivation was to model the bulks as a preparation for modelling of stacks of coated conductors. HTS bulks are usually in the form of solid cylinders. They have therefore a much simpler geometry and structures compared to stacks of HTS coated conductors, which present themselves as cuboids and have laminated layers with different composing materials. As a result, HTS bulks can be described with 2D axisymmetric models, which are also much less complicated in terms of mesh and material types. Based on the developed model, HTS bulks magnetized by PFM using controlled magnetic density distribution coils (CMDCs) are simulated and shown to possibly improve the trapped fields in Chapter 4.2.1. The improvement is not as significant as reported in the literature, which leads to a further discussion about the influence of parameters on the simulation of HTS bulks magnetized by PFM in Chapter 4.2.2. Finally, relevant experiments from the literature are briefly reviewed for evaluating the justifiability and drawbacks of the current model in Chapter 4.3.

4.1 State of the Art

HTS bulks working as trapped-field magnets have always been of interest as potential candidates to replace conventional permanent magnets in a variety of engineering applications. Towards practical application of trapped-field magnets made of bulk materials, effort has been devoted into two directions: improving the material quality to produce high trapped field with good homogeneity and investigating the magnetization techniques to provide reliable, compact and inexpensive magnetization. This thesis is focused on the magnetization of trapped-field magnets. As introduced in Chapter 2.6, the pulsed field magnetization (PFM) method is supposed to be promising to provide *in situ* magnetization of HTS bulks in electrical machines. The research on HTS bulks has a longer history and is thus more extensive than that on stacks of HTS coated conductors, which has become popular only in recent years [68], [73]. In this chapter, the state-of-the-art on the PFM of HTS bulks is briefly reviewed.

The main problem of PFM is that the trapped field is generally lower than that produced by zero field cooling or field cooling, especially at low temperatures. The reason is that the fast dynamic flux motion during PFM causes substantial heat generation, which increases the temperature and decreases the critical current density [2], [71], [78]. The heat generation is understood as a kind of hysteresis loss as described in Chapter 2.5.1. Effort has been devoted to investigate the flux dynamics during PFM and achieve desired trapped fields.

Considering that HTS bulks are usually presented as cylinders, numerical simulations for understanding the flux dynamics are mostly carried out with 2D axisymmetric models. 3D modelling of HTS bulks is mainly for investigating the influence of inhomogeneity resulting from the formation of growth sector boundaries [21], [125], [126]. Considering that the temperature varies greatly during the PFM, the thermal effects are usually considered by including the heat transfer equation as described in Chapter 3.5.

The basic properties of HTS bulks are represented using a critical current density $J_c(B,T)$ which depends on the temperature and the orientation and the magnitude of the magnetic field [28], [29]. Including $J_c(B,T)$ in the numerical simulations of HTS bulks has been a common practice to provide a reliable description of PFM since the temperature and external fields vary greatly during the process [127], [128], [129], [130]. The critical current density $J_c(B,T)$ of bulks can be estimated by measuring the property of a sub-specimen [128], [131], [132] or by deviation from the trapped field profile [133].

The electrical property of HTS bulks is described with the *E*-*J* power law [41] to account for the flux creep during PFM as introduced in Chapter 2.4.2. The exponent parameter n of HTS varies between 5 and 50 [101], based on the strength of the flux creep. The n value is temperature and magnetic field dependent [134], [135]. In [133], the n value is found to be 45 at 77 K after two days by measuring the decay of the remnant magnetization [136], [137]. According to [127], the n value is in the range of 21 to 5 at 77 K for the external field from zero to an arbitrary large value and the n value can be doubled when the temperature is down to 30 K [137].

In literature, however, the exponent parameter n is often supposed to be constant because variation of the n value leads to numerical instability, especially when the thermal influence is considered. Moreover, the n value is difficult to characterize directly for bulks. In [129], the critical state model is used, which corresponds to the situation where the n value approaches infinity. The authors of [75], [83], [128] used the ***A***-*V* formulation with the constant n value of 8 at 40 K. The authors of [85], [86] use the ***H***-formulation and suppose the n value to be 9 during the magnetization and 21 after the magnetization at 40 K. In [16], [21], [45], the authors use the ***H***-formulation while assuming the n value to be 21 for both 40 K and 77 K. The authors of [138] and [139] assume the n value to be 50 at 50 K. The n value influences the strength of flux creep and affects the final trapped fields. The constant n value is acceptable for qualitative investigations of PFM, but for quantitative prediction of experimental results, temperature- and magnetic field-dependent n values should be considered.

The developed models are used to investigate strategies for improving the trapped field produced by PFM. The work on bulks in the literature can be divided into three categories [2]. First are the different magnetization coil arrangements. Using vortex coils instead of common solenoids for magnetization is shown to improve the trapped field by experiments and simulations [74], [75], [129]. The mechanism is explained as reduced heat generation due to a different magnetic flux penetration pattern and better heat transfer along the *ab*-plane of the bulk [76].

Second, the multi-pulse techniques are proved to improve the trapped field produced by PFM effectively by experiments. As introduced in Chapter 2.6.2.2 [78], [79], [80], [81], the techniques include the successive pulsed-field application (SPA), iteratively magnetizing pulsed field method with reducing amplitude (IMRA), and multi-pulse technique with step-wise cooling (MPSC). Numerical simulations of HTS bulks using 2D models suggest that the increase in trapped field by successive pulses is related to the reduction in temperature rise due to the flux already trapped from previous pulses [82], [83].

Third, hybrid or composite structures of HTS bulks may lead to higher or more uniform trapped fields. In [85], [86], [133], HTS bulks with metallic composite structures to improve the thermal conductivity are investigated numerically and experimentally. The research on the influence of composite structures of HTS bulks suggests that stacks of HTS coated conductors as composite bulks have advantages in terms of improved mechanical strength, higher thermal conductivity and more homogeneous critical current density [2].

Inspired by the work on HTS bulks in the literature, the influence of magnetization coils on the PFM of HTS bulks is investigated in Chapter 4.2.1. Then, a discussion on the influence of parameters on the simulation is made in Chapter 4.2.2. Finally, the justifiability and drawbacks of the current model are discussed according to relevant experiments found in the literature in Chapter 4.3.

4.2 2D Modelling

In this chapter, 2D axisymmetric electromagnetic-thermal coupled model based on the ***H***-formulation is used to describe a cylindrical HTS bulk magnetized by pulsed field magnetization. The implementation of the ***H***-formulation is described in Chapter 3.3.2. As a primary stage of the thesis, the assumptions are made consistently with the literature [75], [83], [85], [86], [128], [129]. The magnetic field dependence of the critical current density of the HTS coated conductors is described by the Kim model [28], [29]

$$J_c(B,T) = \frac{J_{c0}(T)}{1+\frac{B}{B_0}}, \tag{4.1}$$

where B is the magnitude of the magnetic field. In the simulations of this chapter, B_0 is assumed to be temperature-independent for simplification and comparison with literature. The temperature dependence of the critical current density is described by [21], [75], [128]

$$J_{c0}(T) = \alpha\left(1-\left(\frac{T}{T_c}\right)^2\right)^{1.5}, \tag{4.2}$$

where the critical temperature T_c is 92 K for YBCO bulk material. The bulk is supposed to be magnetized at 30 K.

4.2.1 Simulation of Pulsed Field Magnetization with Controlled Magnetic Density Distribution Coils

As reported in [74], [75], the trapped field produced by PFM with vortex coils can be improved compared to that with common solenoids. In this chapter, the influence of the coil configurations on the trapped field is further investigated. As shown in Fig. 4.1, three configurations of coils are considered: (a) the common solenoid, which produces a uniform applied field; (b) the vortex coils, which were proved to increase the trapped field; (c) the controlled magnetic density distribution coils (CMDCs), which consist of inner and outer turns carrying currents in opposite directions. For the controlled magnetic density distribution coils, the outer turns carry inversed currents of smaller magnitudes compared to the inner turns; as a result, the generated applied field has a large gradient with the peak value in the center. The ratio between the current density of outer turns and inner turns is defined as α ($0 < \alpha < 1$), which is 0.15 in this work. The applied field distributions generated by the three coils in Fig. 4.1 along the radius of the bulk in the central plane are shown in Fig. 4.2. The idea is to investigate whether the even more non-uniform applied field produced by controlled magnetic density distribution coils can improve the trapped field compared to the vortex coils.

The bulk is cooled from the surface or the sides depending on the position of magnetization coils as shown in Fig. 4.1. A thermal resistive layer of thermal conductivity 0.5 W/(m·K) is placed between the bulk and the boundary of the fixed temperature 30 K [75]. Temperature-dependent heat capacity is taken into account by interpolation of the experimental data, which is approximately 30 to 160 J/(kg·K) between 30 and 80 K. An anisotropic thermal conductivity is considered, which is 4 and 15 W/(m·K) parallel and perpendicular to the axis of the bulk, respectively [76].

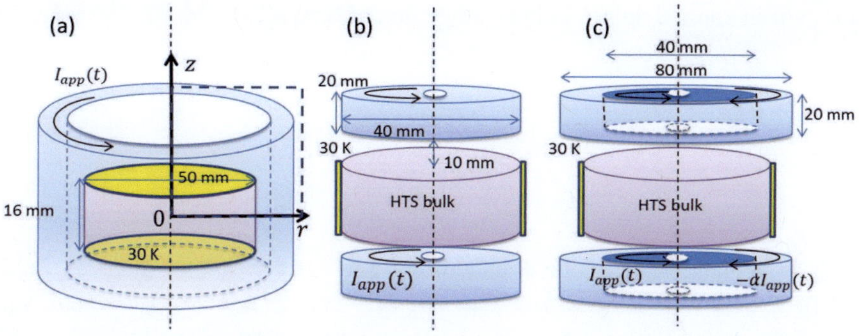

Figure 4.1: Schematics of the model of HTS bulks magnetized by PFM with different coils: (a) solenoid; (b)vortex coils; (c) controlled magnetic density distribution coils (CMDCs), which consist of inner turns and outer turns carrying currents in opposite directions. The ratio between the current density of outer turns and inner turns α $(0 < \alpha < 1)$.

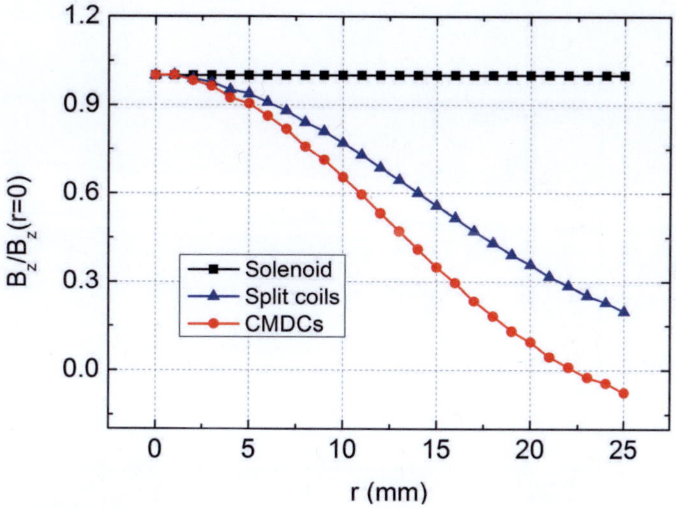

Figure 4.2: The normalized applied magnetic field density distributions generated by the solenoid, split coils and controlled magnetic density distribution coils (CMDCs) in Fig. 4.1 along the radius in the central plane of the bulk.

The normalized applied pulsed field as a function of time is

$$B_{\text{app}} = B_{\text{m}} \sin^2\left(\frac{\pi t}{2\tau}\right), \qquad \text{if } 0 \le t \le \tau \tag{4.3a}$$

$$B_{\text{app}} = B_{\text{m}} \cos^2\left(\frac{\pi(t-\tau)}{10\tau}\right), \qquad \text{if } \tau < t \le 6\tau \tag{4.3b}$$

where B_{m} is the amplitude of the applied pulse and τ is the time of the peak field. The applied field with time is shown in Fig. 4.3. The field ramps up to the peak value in 10 ms and decreases to zero in 50 ms, as an approximation of the typically measured applied field described by Eq. 2.9 [2]. The shape of the pulse is modified in such a way that the initial and ending time derivatives of the function are zero, which improves numerical convergence as mentioned in Fig. 3.1 of [95]. In addition, the modified equation controls the definite end of the applied field. After the applied pulse, the bulk will relax for 30 s to be cooled to 30 K. Consistently with [75], the exponent parameter n in the power law is assumed to be 8, B_0 is 1.3 T in Eq. 4.1. The parameter α is 6.1×10^8 A/m^2 in Eq. 4.2. Considering symmetry, only half of the axisymmetric plane is simulated, as marked by the dashed box in Fig. 4.1.

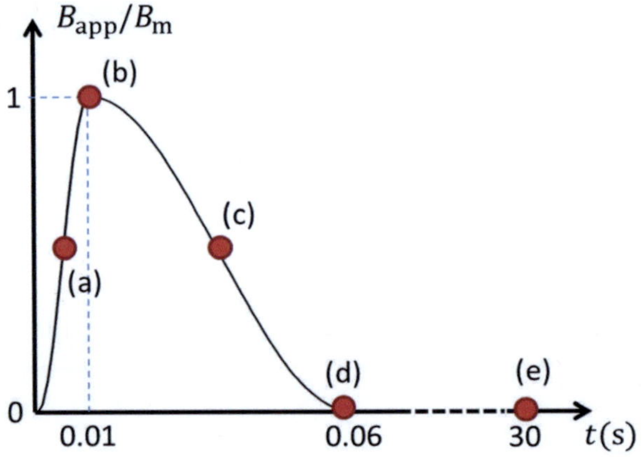

Figure 4.3: The time dependence of the applied field. The marked points are selected points for later analysis: (a) 0.005 s (b) 0.01 s (c) 0.035 s (d) 0.06 s (e) 30 s.

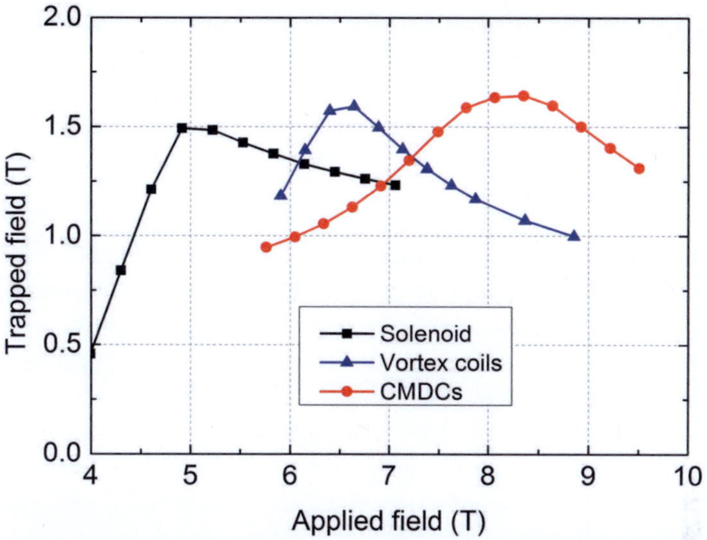

Figure 4.4: The trapped field measured 0.8 mm above the surface center of the bulk with different magnitudes of applied fields produced by the solenoid, split coils and controlled magnetic density distribution coils (CMDCs) in Fig. 4.1 extracted 30 s after the magnetization.

The trapped fields with the amplitudes of the applied fields produced by the three coil arrangements in Fig. 4.1 are compared in Fig. 4.4. The trapped fields are evaluated 0.8 mm above the surface center of the bulk 30 s after the pulse. As shown in Fig. 4.4, for each coil configuration, the trapped field has a maximum value, because a too small applied field is not enough to fully magnetize the bulk, while a too large applied field produces excessive heat and increases the temperature. The results of Fig. 4.2 and Fig. 4.4 show that the larger the spatial gradient of the applied field, the larger the maximum trapped field.

The improvement of the trapped field produced by the vortex coils has been explained as a reduction of produced heat related to the different ways of current penetration [75]. In order to compare the penetration processes by the solenoid, vortex coils and controlled magnetic density distribution coils, the normalized current distributions for the marked points in Fig. 4.3 are compared in Fig. 4.5.

For the solenoid, the currents penetrate from the periphery of the bulk; for the split coils, the currents penetrate more from the surface of the bulk; controlled magnetic density distribution coils push the current penetration to the bulk surface even further. The temperature distributions after the magnetization are compared in Fig. 4.6. The maximum temperatures by controlled magnetic density distribution coils are near the surface of the bulk, which are clearly related to the current distribution in Fig. 4.5: the maximum temperatures appear at the positions where the currents first penetrate the bulk. From time (d) to (e), the currents redistribute due to the flux creep. The maximum temperatures after the magnetization by the vortex coils and controlled magnetic density distribution coils are lower compared to the solenoid, which agrees with the explanation for the improvement in the trapped field in [75].

The simulation agrees with the conclusions reported in the literature that vortex coils can increase the trapped field produced by PFM due to the reduced temperature increase and the trapped field can be even more improved by using controlled magnetic density distribution coils; however, unlike the previous reports, the improvement is not very substantial. In the experiments of [74], the trapped field is improved by 15 %. In [75], the numerical simulation shows 50 % improvement by using the vortex coils. In this work, only less than 10 % improvement by using vortex coils is found. There is the possibility that the trapped field improvement observed in [74] is accidental for the specific sample. In addition, the trapped field recorded in [74] is the maximum field instead of the center field above the bulk, since the sample is not uniform. So far, there are no further experiments comparing different coil configurations using identical samples. The improvement of trapped fields by using vortex coils or controlled magnetic density distribution coils is not fully justified in the view of this thesis.

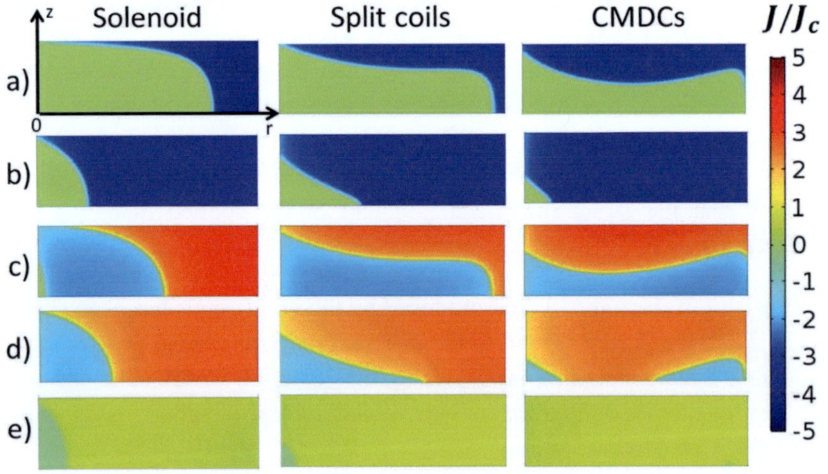

Figure 4.5: Penetration processes during PFM of the solenoid, split coils and controlled magnetic density distribution coils (CMDCs). The applied fields are of the amplitudes when the maximum trapped fields are produced for each coil configuration. (a) to (e) correspond to selected time points in Fig. 4.3. Only half of the axisymmetric plane is plotted as marked in Fig. 4.1.

The discrepancy between the simulations in this thesis and the results reported in [75] could be related to the values of some simulation parameters. There are many differences between the simulations in this chapter and in [75]. For example, in [75] the heat capacity is assumed to be constant as 132 J/(kg·K); in this work a temperature-dependent heat capacity, which is 30 to 160 J/(kg·K) between 30 and 80 K, is used. This may result in larger temperature increase and less trapped field. The critical electrical field E_c for the definition of the critical current density is 1×10^{-6} V/m in [75]; in this thesis E_c is 1×10^{-4} V/m according to the common approach for HTS. The position for extracting the trapped fields is right on the surface of the bulk in [75], while in this thesis it is 0.8 mm above the surface, which is a practically measurable position; the extracted trapped field above the surface can be 20% less than that on the bulk surface in this geometry. Other important parameters include the exponent parameter n in the power law and the field parameter B_0 in Eq. 4.1, which are difficult to be characterized for bulks. Ideally the temperature dependence of such parameters should be considered in the simulation. Following the discussions in this chapter, the influence of parameters on the simulations of HTS bulks magnetized by PFM is further discussed in Chapter 4.2.2.

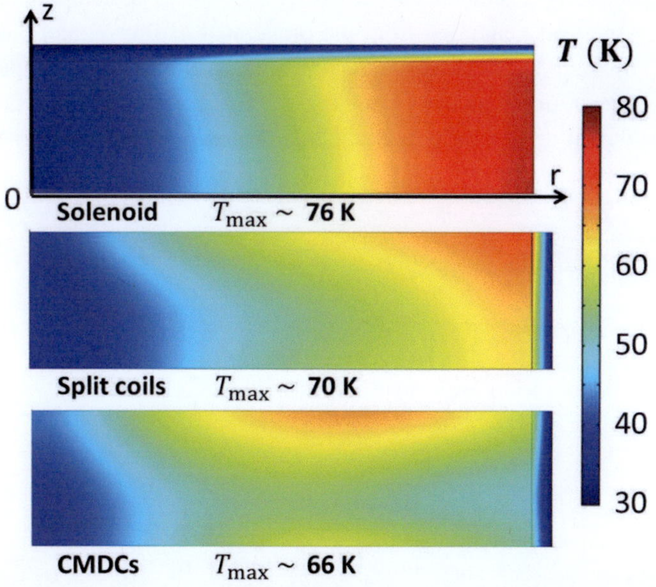

Figure 4.6: The temperature distribution after the pulse (t=0.06 s) for the solenoid, split coils and controlled magnetic density distribution coils (CMDCs), respectively. The applied fields are of the amplitudes when the maximum trapped fields are produced for each coil configuration. Only half of the axisymmetric plane is plotted as marked in Fig. 4.1.

4.2.2 Influence of Parameters on the Simulation

In this chapter, the influences of parameters to the simulation, including the factor n in the E-J power law and B_0 in the Kim model in Eq. 4.1, are compared with both the electromagnetic-thermal coupled model and the electromagnetic model. The modelling scheme is the same as described in the previous Chapter 4.2.1. The time dependence of the pulsed applied field is shown in Fig. 4.7. An HTS bulk magnetized by PFM with a common solenoid is considered as shown in Fig. 4.8. Considering symmetry, only half of the axisymmetric plane is simulated.

The cooling of the bulk is represented by an average heat transfer flux ϕ_{cool} on the top surface of the bulk according to [86]

$$\phi_{\mathrm{cool}} = \frac{1}{A} \times \begin{cases} 0, & \text{if } T < 30\ \mathrm{K} \\ (T-30)\ \ \mathrm{W/m^2}, & \text{if } 30\ \mathrm{K} \leq T \leq 45\ \mathrm{K}, \\ 15\ \mathrm{W/m^2}, & \text{if } T > 45\ \mathrm{K} \end{cases} \tag{4.4}$$

where A is the top surface area of the bulk. All the other boundaries of the simulated geometry are thermally insulated as shown in Fig. 4.8. The temperature of the bulk increases due to the heat generated from the movement of flux in and out of the sample during PFM. The temperature then decreases back to 30 K after the pulse due to the cooling. The temperature-dependent heat capacity is interpolated from the experimental data [76]. The thermal conductivity is 3 and 20 W/(m·K) parallel and perpendicular to the axis of the bulk, respectively [76].

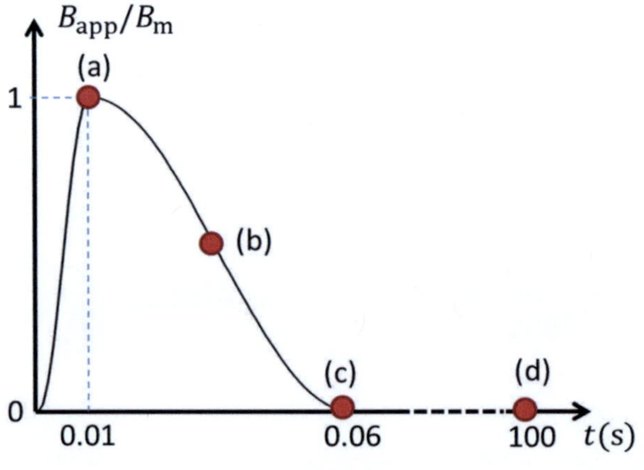

Figure 4.7: The time dependence of the applied field. The marked points are selected points for later analysis: (a) 0.01 s (b) 0.035 s (c) 0.06 s (d) 100 s.

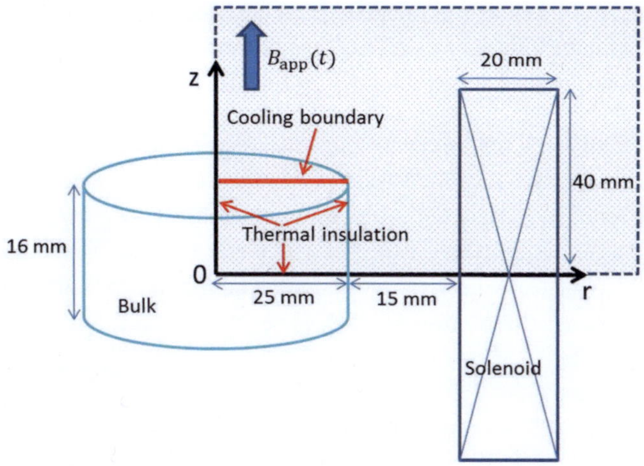

Figure 4.8: The geometry of the model. Only half of the axisymmetric plane (shaded area) is modelled. The plot is not to scale. The upper boundary of the bulk is set as a cooling boundary with a given heat flux. The other three boundaries of the bulk are thermally insulated. See text for details.

First, the influence of the exponent parameter *n* in the *E-J* power law is investigated. The parameter B_0 for Eq. 4.1 is fixed as 1.3 T. For the given geometry, the trapped field by field cooling is 2.34 T by assuming that the current density is the critical current density everywhere inside the bulk [140]. This value is estimated by the critical state model assumption, and can be higher than what is measured in experiments. Due to flux creep [41], [133], [135] the electrical field in the bulk after magnetization is probably lower than the presumed E_c of 1×10^{-4} V/m. Correspondingly, the remnant currents are lower than the critical current density.

The *n* values of 8, 21 and 50, which were often used in the literature [16], [21], [45], [75], [83], [84], [85], [86], [128], [138], [139], are compared. The trapped fields and the maximum temperatures after the magnetization with the amplitudes of the applied fields calculated with the electromagnetic-thermal coupled model and the pure electromagnetic model are shown in Fig. 4.9. The trapped fields are taken 0.8 mm above the surface center of the bulk 100 s after the magnetization.

For the trapped field calculated by electromagnetic-thermal models, there is an optimal applied field, which produces a maximum trapped field, because too small applied fields are not enough to fully magnetize the bulk, but too large applied fields will generate excessive heat and increase the temperature. The maximum temperature in the bulk also shows that the temperature increases with the increasing amplitude of the applied field. For the trapped fields with electromagnetic models, the trapped field increases with the amplitude of the applied field and then becomes constant, because the temperature rise is not considered. For the simulation of PFM, the thermal effects have to be considered, so the electromagnetic-thermal coupled model is the valid approach [2].

Comparing different *n* values in electromagnetic-thermal models, the maximum trapped field and the corresponding applied field are larger if the *n* value is larger. Smaller *n* values result in more over critical current and heat generation during the magnetization, which lead to larger temperature rise. For the electromagnetic models, the maximum trapped field is also larger if the *n* value is larger. The flux creep is stronger and the decay rate is higher if the *n* value is lower [135], [136], [137], so the trapped is lower at the same reference time ($t = 100$ s).

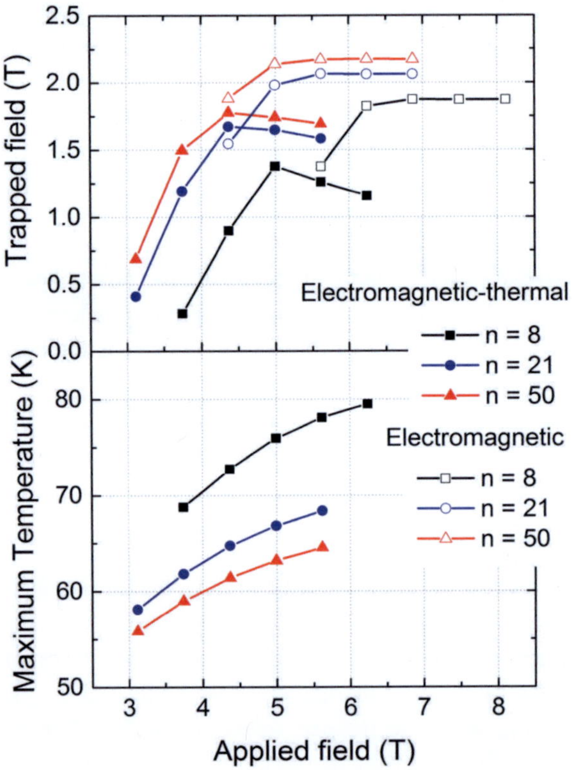

Figure 4.9: The trapped field taken 0.8 mm above the surface center of the bulk and the maximum temperature with different magnitudes of applied fields after relaxing ($t = 100$ s). Different *n* values in the *E-J* power law are compared. For the electromagnetic model, the temperature is constant at 30 K.

When the applied field is 5.6 T, the time evolution of the trapped fields and the maximum temperature calculated with the electromagnetic-thermal coupled model (EMT) and the electromagnetic model (EM) are compared in Fig. 4.10. The normalized current distributions J/J_c at different times are shown in Fig. 4.11.

From 0 s to 0.01 s, the applied field ramps up and the currents penetrate from the periphery of the bulk. For either electromagnetic or electromagnetic-thermal model, the current penetration for *n* of 50 is deeper than that for *n* of 8. When *n* is 8, the current density is much higher than the critical current density. In Fig. 4.9, the magnetic fields at 0.01 s are smaller when the *n* value is smaller, because the smaller *n* value results in larger screening fields with more over critical currents. For each *n* value, the current penetration of the electromagnetic-thermal model is deeper than that of the electromagnetic model, because the critical current density decreases as the temperature increases when thermal effects are considered. From 0.01 s to 0.06 s, the applied field decreases to zero and currents in opposite directions start to penetrate from the periphery of the bulk. After 0.06 s, the applied field is zero, but the current distribution keeps changing due to the flux motion described by the power law [135], [136], [137].

In the electromagnetic-thermal models, the maximum temperature rises quickly from 0 to 0.01 s during the ramping period of the applied field. Then the maximum temperature rises more slowly from 0.01 s to 0.06 s during the damping period of the applied field. There are two reasons for this: first, the varying rate of the field damping is slower than that of the field ramping; second, the heat capacity of the bulk is larger at higher temperatures. Finally, the temperature increase for n=8 is larger than that for n= 50, because the former produces more over critical currents during the magnetization. From 0.06 s to 100 s, the rate of decay of the magnetic field keeps decreasing, because the temperature of the bulk is decreasing and the critical current density is increasing.

In the model, the normalized current density J/J_c can be much larger than 1 (almost 5 when n=8) during the magnetization. The solution mathematically matches Maxwell's equations and the non-linear *E-J* assumption; however, whether it makes sense physically is unclear. The *E-J* power law is not going to hold infinitely for over-critical current density [43], [44], [45], [141], [142], [143]. However, a precise *E-J* relationship for the HTS materials in over critical regions, especially for HTS bulks, is not available yet. Different *n* values lead to different dynamics during the process. To use small *n* values for the simulation of PFM as frequently reported in the literature seems doubtful.

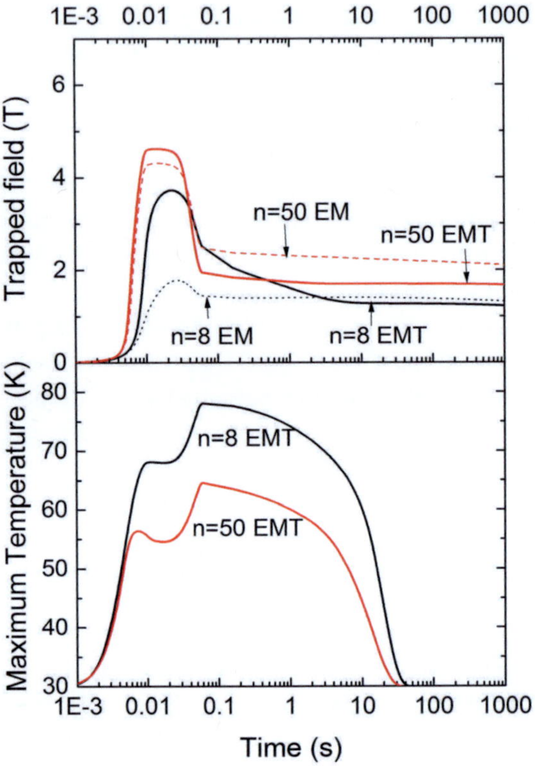

Figure 4.10: The magnetic field measured 0.8 mm above the surface center of the bulk and the maximum temperature with time. The applied field is 5.6 T. EMT: electromagnetic-thermal coupled model; EM: pure electromagnetic model. For the electromagnetic model, the temperature is constant at 30 K.

When the *n* value is fixed as 21, three different B_0 values, 0.3 T, 1.3 T and 4 T in Eq. 4.1 are compared. The corresponding trapped fields produced by field cooling are estimated to be 1.30 T, 2.34 T and 3.37 T by assuming that the current density is the critical current density everywhere in the bulk [140]. The trapped fields and the maximum temperature during the magnetization with the amplitudes of the applied fields calculated with the electromagnetic-thermal coupled model and the electromagnetic model are shown in Fig. 4.12.

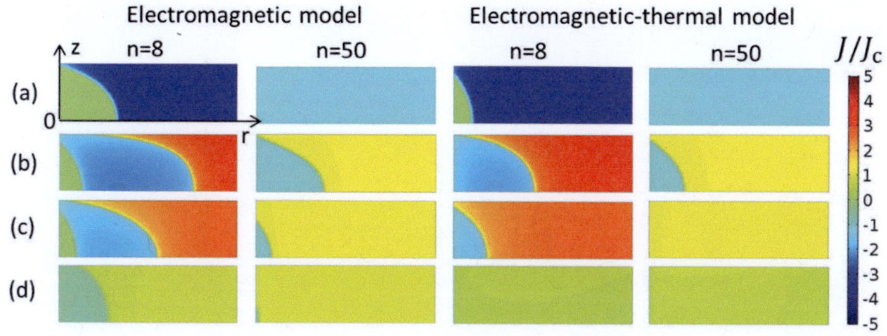

Figure 4.11: The normalized current density distributions $\boldsymbol{J}/\boldsymbol{J}_\mathbf{c}$ at different times as marked in Fig. 4.7. Only half of the axisymmetric plane of the bulk is plotted as marked in Fig. 4.8. The applied field is 5.6 T.

For the electromagnetic-thermal model and the electromagnetic model, the larger the B_0 value, the larger the optimal applied fields. For the electromagnetic-thermal model, the maximum trapped fields increase non-substantially when the B_0 value is increased from 1.3 to 4 T. The ratio between the maximum trapped field produced by PFM and the trapped fields produced by field cooling is 84 %, 71 % and 52 % when the B_0 value is 0.3 T, 1.3 T and 4 T, respectively. The reason is that the bulk of larger B_0 value will generate more heat during the magnetization. This can be further demonstrated by the fact that the larger B_0 values result in higher maximum temperatures. For the PFM process when the temperature rises greatly, the B_0 value greatly influences the trend of the trapped field with the applied field, which suggests that taking into account the temperature dependence of B_0 is necessary.

In summary, the material parameters used for simulations of HTS bulks magnetized by PFM can greatly influence the simulation results. Smaller n values in the E-J power law can result in a larger temperature increase and stronger current decay after the magnetization. The maximum trapped field does not increase monotonically with the B_0 parameter in the Kim model. Since the simulation results strongly depend on the selected parameters, ideally the temperature dependence of the n value and the B_0 value should be considered in the simulation of HTS bulks. The results in Chapter 4.2.2 are published in [145].

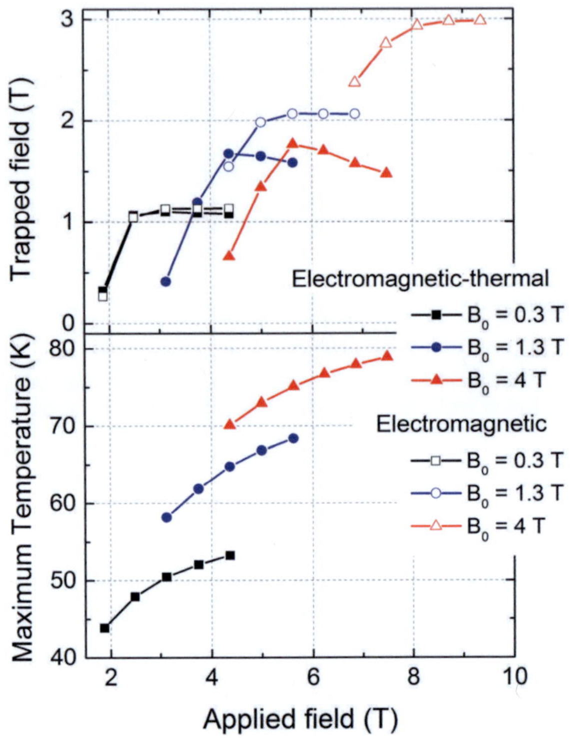

Figure 4.12: The trapped field taken 0.8 mm above the surface center of the bulk and the maximum temperature with different amplitudes of applied fields after relaxing (t = 100 s). Different B_0 values in the Kim model in Eq. 4.1 are compared. For the electromagnetic model, the temperature is constant at 30 K.

4.3 Experimental Evidence in the Literature

The material properties of HTS bulks are difficult to reliably characterize. Moreover, most HTS bulks are actually very inhomogeneous. During the PFM process, the temperature varies greatly and changes the material properties; besides, the cooling condition is difficult to judge. There are no quantitative simulations of HTS bulks magnetized by PFM that properly match experiments in the literature. For the field cooling process, there are some quantitative simulations when the sample is relatively homogeneous [16], [144].

For most simulations of PFM, the target is not to make quantitative predictions for the experiments, but to understand the mechanism of PFM and to optimize the process. Since the parameters greatly influence the simulation results [145], they should be considered carefully even for qualitative simulations. Based on the simulations in this chapter, two indicators can be considered to assess the simulation: the ratio between the applied and trapped field, and the ratio between the trapped field produced by PFM and that by field cooling.

The ratio between the applied and the trapped field reflects the strength of flux creep and flow during the PFM. For the ideal zero field cooling, which corresponds to no flux creep, this ratio should be 2 when the external field is just enough for the magnetization [2]. In reality, the trapped field can only be measured at the surface instead of in the center of the sample, so this ratio is a bit larger than 2, depending on the geometry. For the PFM, the situation is much more complicated. As discussed in Chapter 4.2.2, there is an optimal applied field which produces the maximum trapped field. The ratios between the optimal applied fields and the maximum trapped fields are 3.6, 2.6 and 2.5 when the n values are 8, 21 and 50, respectively, as shown in Fig. 4.9. The smaller n value suggests stronger flux flow, which leads to a larger ratio of the applied field and the trapped field.

In the experiments presented in the literature, the ratio between the applied field and the trapped field is 2.6 in [146] and 5.4 in [147] at 77 K. For the same sample in [148], the ratio is 5.3 at 70 K, 4.5 at 50 K and 4.8 at 30 K. In [149], the ratio is 4 at 77 K and 3 at 35 K. In [21], the ratio is 5.3 at 65 K and 4.3 at 40 K for an YBCO bulk, and 4.6 at 65 K and 3.8 at 40 K for a homogeneous GdBCO bulk. Most experiments of PFM reveal that the HTS bulks behave as if the flux creep is quite strong during the PFM, which corresponds to a low n value around 8 based on the simulations in this chapter; however, the n values of HTS bulks measured by observing the rate of decay of the remnant magnetization are between 20 and 50 at 77 K and are much larger at lower temperatures [133], [136], [137]. In the view of this thesis, this discrepancy is related to the inhomogeneity of HTS bulks. The inhomogeneity and growth sector boundaries can completely change the current density distributions and the ways of penetration [146], [147], [148], [149] and extra heat can be generated [21], which cannot be properly described by the current 2D axisymmetric models. Besides, flux jumps are not properly described by the current models and may be another reason for the discrepancy between simulations and experiments. Further research is needed for reasonable simulations of HTS bulks.

The ratio between the trapped field produced by PFM and that produced by field cooling reflects the loss of trapped field due to heat generation caused by the fast flux motion during PFM. The trapped field produced by field cooling is the maximum possible trapped field of a given sample [2]. Based on the simulations in Chapter 4.2.2, smaller n

values in the power law and larger B_0 values in the Kim model can lead to smaller ratios between the trapped fields produced by PFM and those by field cooling. In [128], it is shown that high critical current density can result in lower trapped field produced by PFM. The B_0 values in the Kim model and the critical current density are larger at lower temperatures. Experiments in the literature reveal that the trapped field of HTS bulks by PFM cannot be improved much by going to lower temperatures [73], which means that the ratio between the trapped field produced by PFM and that by field cooling is smaller at lower temperatures. The current 2D models in the literature can already qualitatively reproduce this experimental observation [2].

Based on the simulation results presented in this chapter, the next chapter focuses on the simulations of stacks of HTS coated conductors magnetized by PFM. Compared with HTS bulks, HTS coated conductors have more complicated laminated structures of different materials; however, they are more uniform and can be characterized reliably, which makes simulations reliable, both qualitatively and quantitatively. Temperature- and magnetic field-dependent critical current density and temperature-dependent material properties of other composing materials are considered properly based on experimental data. In order to facilitate the convergence of the simulations, the temperature and magnetic field dependences of the n value in the E-J power law are not considered. The simulations are validated by experiments qualitatively and some specific simulations agree with the experiments quantitatively.

5 Magnetization of Stacks of HTS Coated Conductors

Stacks of HTS coated conductors can work as composite HTS bulks to become trapped-field magnets after magnetization. Compared to pure HTS bulks, stacks of coated conductors have several advantages: the HTS coated conductors are more commercially available and the stacks are relatively cheaper; the mechanical strength of stacks is better due to the metallic substrates; and the shapes of stacks are more flexible to adapt to different configurations of applications [49]. Furthermore, HTS tapes are easier to characterize reliably to allow reasonable modeling; and they are more uniform without obvious growth sector boundaries as in bulks [21], [125], [126]. Considering that the nonlinear *E–J* properties of bulks and tapes are generally similar, the general conclusions of this chapter may apply to HTS bulks as well.

In this chapter, simulations of HTS coated conductors magnetized by pulsed field magnetization (PFM) are first carried out with 2D electromagnetic-thermal coupled models. Based on the simulations, two strategies are proposed to improve the trapped fields produced by PFM. The 2D modelling is then extended to 3D for a common cuboid flat stack and a curved stack. In the end, experiments are carried out to validate the simulations.

5.1 2D Modelling

In this chapter, a 2D planar electromagnetic-thermal model is developed to describe an infinitely long stack of coated conductors magnetized by PFM. The model considers temperature and anisotropic magnetic field-dependent critical current density $J_c(\boldsymbol{B},T)$ of HTS. Other temperature-dependent thermal and electrical material properties of all composing materials are also considered including the heat capacity c_p, the thermal conductivity λ, and the electrical resistivity ρ_e. Based on the model, two strategies are suggested to improve the trapped field of a stack produced by PFM.

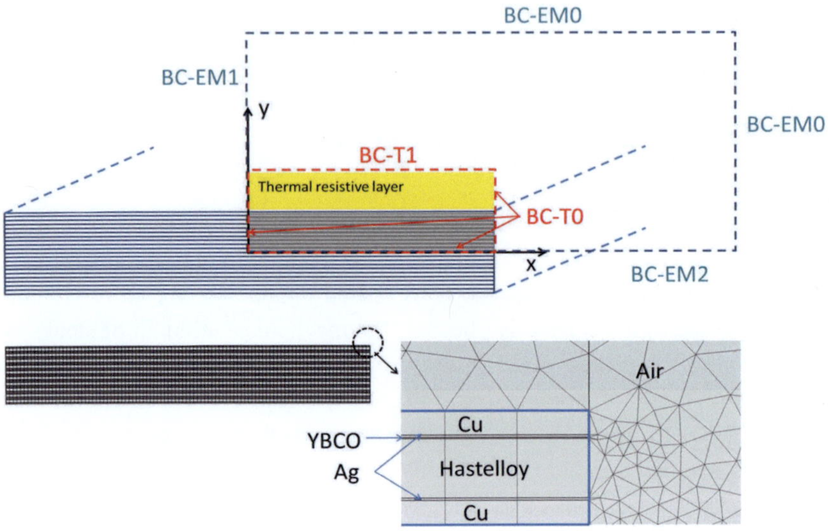

Figure 5.1: The geometry of the cross-section of a stack of 20-layer coated conductors used for 2D simulation. Only the first quarter of the cross-section is simulated to save computation time. BC-EM0, 1 and 2 are boundaries for the electromagnetic equations and BC-T0 and 1 are the boundaries of the heat equation. The stack meshed with mapped mesh is shown on bottom left. The domain around the stack is meshed with triangular elements as zoomed-in on bottom right. The realistic laminated structure of HTS tapes according to Fig. 2.2 is considered including all composing materials except the buffer layer.

5.1.1 Geometry, Mesh and Boundary Conditions

The model is 2D, accounting for an infinitely long stack based on the ***H***-formulation of Maxwell's equations and heat transfer equation implemented in COMSOL as described in Chapter 3.3.1. The tapes are assumed to be YBCO tapes (SCS4050) manufactured by SuperPower Inc. [26], [27]. The composition of the tape is shown in Fig. 2.2, including the HTS material, silver cover layers, copper stabilizers and the Hastelloy substrate. The realistic laminated structure of the HTS tape is considered except the buffer layers, which are only 0.2 µm thick resulting in no physical effects [84].

The stack is composed of 20 superposed 12 mm-wide tapes. Considering symmetry, only one quarter of the geometry is simulated to save computation time, as shown in Fig. 5.1. The thickness of the different layers ranges from 1 to 50 µm, which is rather small compared to the width of the tape. Considering the extremely large aspect ratio, the HTS coated conductors are meshed with mapped elements as shown at the bottom left of Fig. 5.1. There is one division along the thickness of each material and 100 divisions along the half width of the tape. The domain around the stack is meshed with free triangular elements as zoomed-in at the bottom right of Fig. 5.1.

Considering symmetry, the boundary conditions for electromagnetic equations are magnetic flux parallel to BC-EM1 and magnetic flux perpendicular to BC-EM2 in Fig. 5.1. On BC-EM0 lines, which are far away from the centered stack, the magnetic flux density is prescribed according to the uniform external applied magnetic field.

The applied field for the PFM is given by

$$B_{\mathrm{app}} = B_{\mathrm{m}} \sin^2\left(\frac{\pi t}{2\tau}\right), \qquad \text{if } 0 \leq t \leq \tau \tag{5.1a}$$

$$B_{\mathrm{app}} = B_{\mathrm{m}} \cos^2\left(\frac{\pi(t-\tau)}{10\tau}\right), \qquad \text{if } \tau < t \leq 6\tau \tag{5.1b}$$

where τ is the characteristic time and B_{m} is the field amplitude. Similar to the approach in Chapter 4.2, the equation is an approximation of Eq. 2.9. The applied field ramps from zero to the peak value quickly and goes back to zero slowly. The equation is modified so that the initial and ending time derivatives of the function are zero, which improves numerical convergence. In the simulation of Chapter 5.1, the characteristic time τ is set equal to 10 ms.

The domain for the heat equation is smaller than that of the electromagnetic equations. The corresponding boundaries are marked with the red dashed red lines in Fig. 5.1. A thermal resistive layer separates the boundary and the tape surface. The layer is 1 mm thick with thermal conductivity of 0.1 W/(m·K), and can be adjusted to describe non-ideal cooling [75], [83]. A Dirichlet boundary condition is used for BC-T1 to set the equilibrium temperature T_0 at 30 K. On BC-T0, thermal insulation is assumed. With these boundary conditions, the stack is cooled only when its temperature is above T_0.

5.1.2 Material Properties

Comprehensive temperature-dependent material properties are considered, including the critical current density J_{c} of HTS, the electrical resistivity ρ_{e}, the heat capacity c_p and the thermal conductivity λ of all composing materials. The thermal contact between tapes is considered to be perfect.

The critical current density J_{c} of HTS coated conductors depends on the temperature and the magnitude and orientation of local magnetic field. As mentioned in Chapter 2.3, the dependence on magnetic field is typically anisotropic for coated conductors. In order to make reasonable assumptions for the critical current density J_{c}, the lift factor measurement data of YBCO superconducting tapes from SuperPower is used [150]. The lift factor L is defined as the ratio between the critical current in an external field at a lower temperature $I_c(\boldsymbol{B},T)$ and the self-field critical current of the tape $I_{c0}(\text{self}, 77\text{ K})$ at 77 K.

$$L(\boldsymbol{B},T) = \frac{I_c(\boldsymbol{B},T)}{I_{c0}(\text{self}, 77\text{ K})}. \quad \textbf{(5.2)}$$

The critical current density J_c is obtained by dividing the measured critical current I_c by the cross-section area A of the HTS layer. This is considered to be a valid approach, because the PFM works in strong magnetic field and the influence of the self-field can be neglected. Parameters for each temperature are fitted with the equation with the least root mean square value method

$$L(\boldsymbol{B},T) = \frac{J_c(\boldsymbol{B},T)}{J_{c0}(\text{self}, 77\text{ K})} = \frac{l_{c0}(T)}{\left(1 + \frac{\sqrt{(k(T)B_x)^2 + B_y^2}}{B_0(T)}\right)^{\beta(T)}}, \quad \textbf{(5.3)}$$

where B_x and B_y are the components of the magnetic flux density parallel and perpendicular to the flat face of the tape, respectively, and $l_{c0}(T)$, $k(T)$, $B_0(T)$ and $\beta(T)$ are temperature-dependent parameters. The parameters obtained for each temperature are listed in Tab. 5.1. Each parameter varies smoothly and monotonically with the temperature, so parameters for arbitrary temperatures are estimated by directly interpolating between temperatures. The measured lift factors L and the fitted ones with parallel field B_x and perpendicular field B_y for given temperatures are shown in Fig. 5.2. In this work, J_c(self, 77 K) is assumed to be 3 MA/cm^2.

Table 5.1: Fitted parameters of the lift factor of a SuperPower tape with Eq. 5.3 for given temperatures

T (K)	l_{c0}	k	B_0 (T)	β
20	6.52	0.06	5.04	1.64
30	6.12	0.07	3.23	1.41
40	5.29	0.10	1.86	1.12
50	4.12	0.17	1.26	0.96
65	2.44	0.61	0.59	0.77

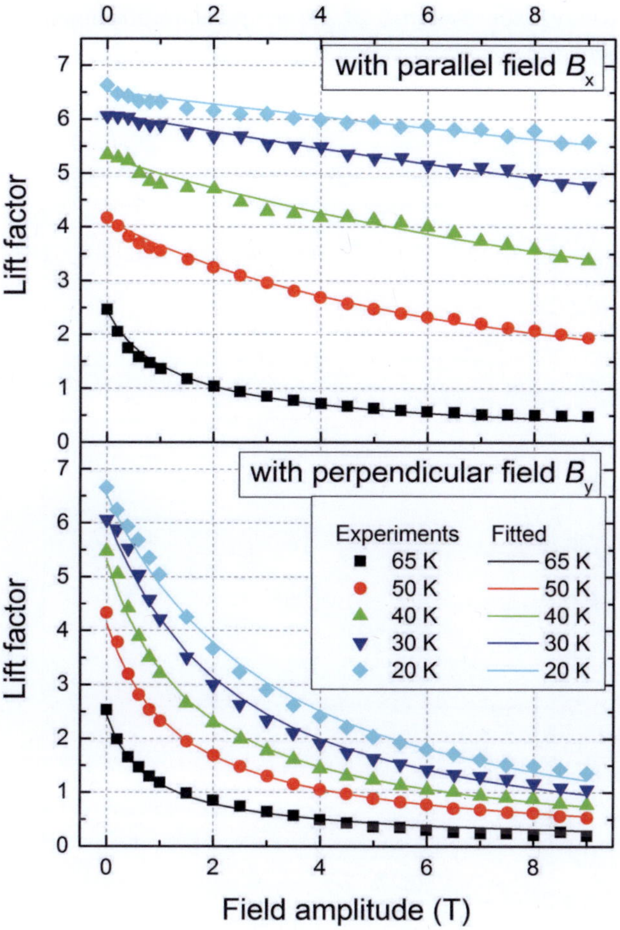

Figure 5.2: The measured lift factors and the fitted lift factors of a SuperPower tape using Eq. 5.3 with parameters listed in Tab. 5.1 for given temperatures. The upper graph and lower graph show the lift factor with parallel field B_x and perpendicular field B_y, respectively [150].

Other temperature-dependent material properties, including electrical resistivity ρ_e, heat capacity c_p and thermal conductivity λ are also considered in this model by directly using interpolated data from experiments [76], [151], [152], [153], [154], [155]. The used values are shown in Fig. 5.3, which can be downloaded from [156], [157]. The residual resistance ratio (RRR) is 42 for copper as suggested by [152] for tapes produced by SuperPower. The temperature dependence of the heat capacity c_p is especially important because it avoids unrealistic over-rise of the temperature and improves the numerical stability of the model.

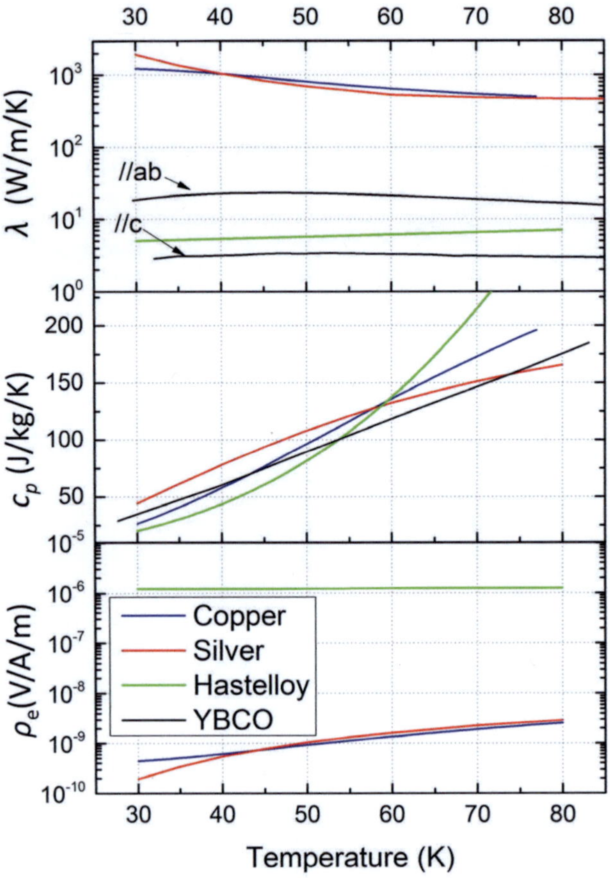

Figure 5.3: The temperature-dependent properties of composing materials of the SuperPower HTS tape in the 2D simulation, including thermal conductivity λ, heat capacity c_p and electrical resistivity ρ_e. The anisotropic thermal conductivity of YBCO along different directions is shown, respectively [157].

5.1.3 Simulation of Pulsed Field Magnetization with Controlled Magnetic Distribution Coils

As mentioned in Chapter 4.2.1, magnetization of HTS bulks by PFM with vortex coils instead of a common solenoid has been attracting interest. Stacks of HTS coated conductors are generally similar to HTS bulks when working as trapped-field magnets since they both contain HTS materials that obey the *E-J* power law; however, the laminated metals present in HTS stacks can generate eddy current loss and also influence heat diffusion, which result in different behaviors during magnetization. In this chapter, the use of controlled magnetic density distribution coils (CMDCs) is extended to HTS stacks. The roles of different materials in HTS stacks are discussed as well. The results in Chapter 5.1.3 are published in [157].

The controlled magnetic density distribution coils consist of a double pair of vortex coils as shown in Fig. 5.4(b). Each pair of vortex coils consists of two coils carrying inverse currents. The inner coil with current I_1 generates the main applied field. The outer coil carries an inverse current $I_2 = -\alpha I_1$ $(0 < \alpha < 1)$ which weakens the applied field on the periphery of the stack. The controlled magnetic density distribution coils thus can generate a non-uniform field along the x axis, with a peak in the center. By adjusting α, the gradient of the applied field can be adjusted. As shown in Fig. 5.5, three controlled magnetic density distribution coils generate non-uniform magnetic field compared to the common solenoid. The controlled magnetic density distribution coils produce a low-, medium- and high-gradient applied field with α value of 0.6, 0.62 and 0.64, which are named as CMDC-low, CMDC-medium and CMDC-high, respectively.

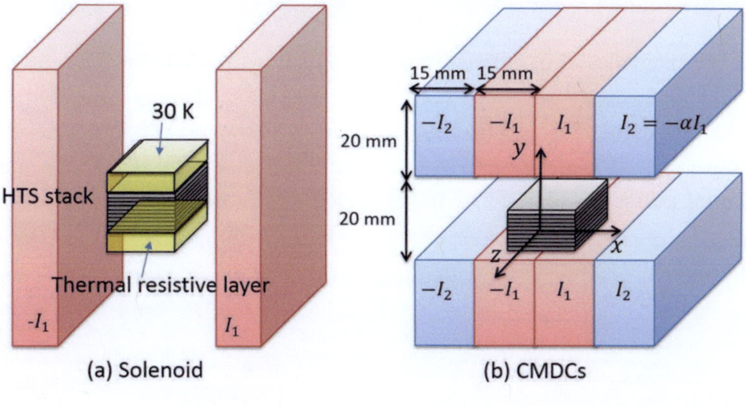

Figure 5.4: Schematics of the coils used for the PFM of HTS stacks. (a) Solenoid; (b) Controlled magnetic density distribution coils (CMDCs). The plots are not to scale.

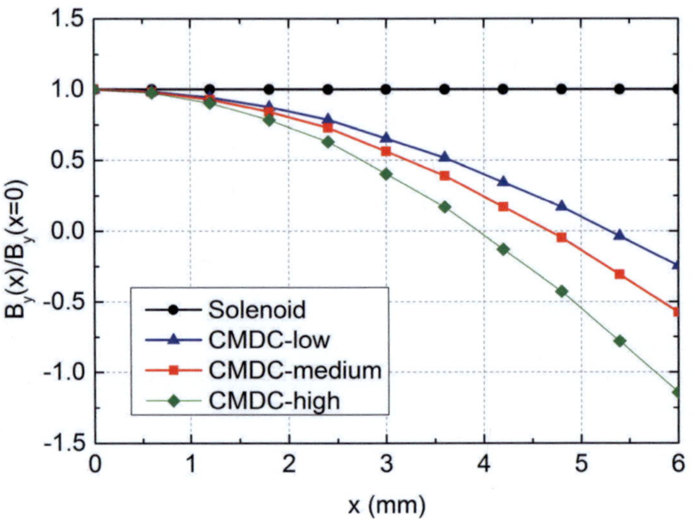

Figure 5.5: The magnetic field density distributions of the applied field produced by different coils in Fig. 5.4 along the tape width (from x = 0 to 6 mm). The solenoid produces a uniform field. Fields with different gradients are produced by different controlled magnetic density distribution coils, namely CMDC-low, CMDC-medium and CMDC-high, corresponding to α values of 0.6, 0.62 and 0.64, respectively.

As described in Chapter 5.1.1, only one quarter of the geometry is simulated. The cooling boundary condition is prescribed by putting a thermal resistive layer between the stack and a constant temperature 30 K as shown in Fig. 5.4, which applies to both the solenoid and controlled magnetic density distribution coils.

The stacks are magnetized by pulsed magnetic fields with different magnitudes using the solenoid and the three different controlled magnetic density distribution coils as described in Fig. 5.4 and 5.5. The trapped field is extracted 0.8 mm above the surface center of the stack. It is evaluated 10 s after the pulse, which allows the stack to fully relax and cool down to 30 K. The trapped flux is defined as the integration of the y component of the remnant field B_y from x = 0 to 6 mm. The trapped field and flux obtained by the solenoid and controlled magnetic density distribution coils are compared in Fig. 5.6.

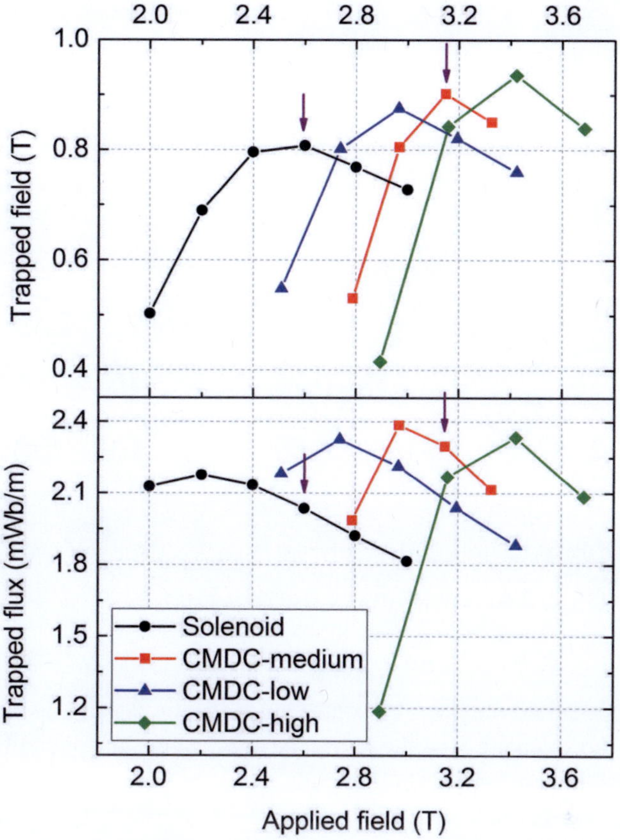

Figure 5.6: The trapped field extracted 0.8 mm above the surface center of the stack 10 s after the applied pulsed field of different amplitudes. The trapped flux defined as the integration of the field from x = 0 to 6 mm. The four lines show the results obtained by using different coils as described in Fig. 5.4 and 5.5. The marked points will be used for later analysis.

For each arrangement of coils, the trapped field first increases and then decreases with the applied field. This is consistent with numerous previous theoretical and experimental works on HTS bulks [2], [74], [75], [78], [82], [83], [85], [86], [148], [149]. A too small applied field cannot induce full currents in the sample; however, a too large applied field generates excessive heat, increasing the temperature T and reducing the critical current J_c. There is an optimal applied field, which results in the maximum trapped field.

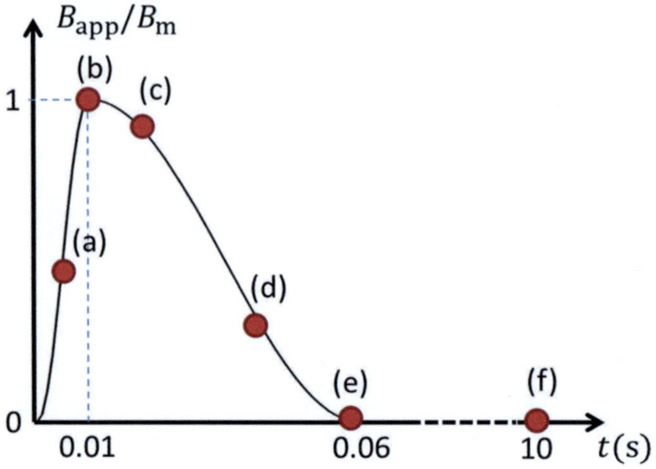

Figure 5.7: The shape of the applied pulsed field as described by Eq. 5.1. The marked time points are used for presenting the current distribution during the magnetization for Fig. 5.6. Selected points for analysis are: (a) 0.005 s (b) 0.01 s (c) 0.015 s (d) 0.04 s (e) 0.06 s (f) 10 s.

The maximum trapped field obtained by controlled magnetic density distribution coils is larger than that by the solenoid. And the larger the gradient of the applied field generated by controlled magnetic density distribution coils, the larger the maximum trapped field. This finding shows consistency with [74], [75], which found that the vortex coils can generate larger trapped field in HTS bulks compared to solenoids.

The penetration processes during PFM by solenoid and controlled magnetic density distribution coils with medium gradient of applied field (CMDC-medium) at time points as marked in Fig. 5.7 are compared in Fig. 5.8. The normalized critical current density J/J_c is shown with the color scale. The applied fields in this figure are of magnitudes 2.6 T for the solenoid and 3.2 T for the CMDC-medium when the maximum trapped fields are obtained as pointed by the arrows in Fig. 5.6.

As shown at time point (f) in Fig. 5.7 and Fig. 5.8, the maximum trapped field is acquired when the stack is not fully filled with positive currents. This suggests that for PFM at low temperature, using the full cross-section of the sample to carry remnant currents generates excessive heat and has negative influence on producing more trapped fields. At time point (f), J/J_c is only around 0.5, suggesting a loss of trapped field due to dynamics during PFM compared to field cooling. From time points (e) to (f), currents decay quickly and then stabilize. At the same time, the critical current density J_c increases when the temperature recovers to 30 K.

Comparing the two coils, the penetration of controlled magnetic density distribution coils starts from the surface of the stack and the current front tends to be parallel to the surface, unlike that of the solenoid. The penetration of controlled magnetic density distribution coils results in less heat generation on the periphery of the stack compared to the solenoid situation.

The magnetic field and flux lines distribution at time point (a) are shown in Fig. 5.9. For the solenoid, the flux lines tend to accumulate on the periphery of the stack; for the controlled magnetic density distribution coils, the flux lines arrange parallel to the stack surface and the magnitude of the magnetic fields inside the stack is lower. Moreover, the field in the controlled magnetic density distribution coils situation is mainly in the direction parallel to the flat surface of the tape. In this way, the critical current density J_c is less degraded according to Eq. 5.3 in the controlled magnetic density distribution coils situation. As a result, controlled magnetic density distribution coils need higher field to penetrate the stack compared to the solenoid. The less degraded critical current density J_c by controlled magnetic density distribution coils partially explains why the final trapped field can be higher.

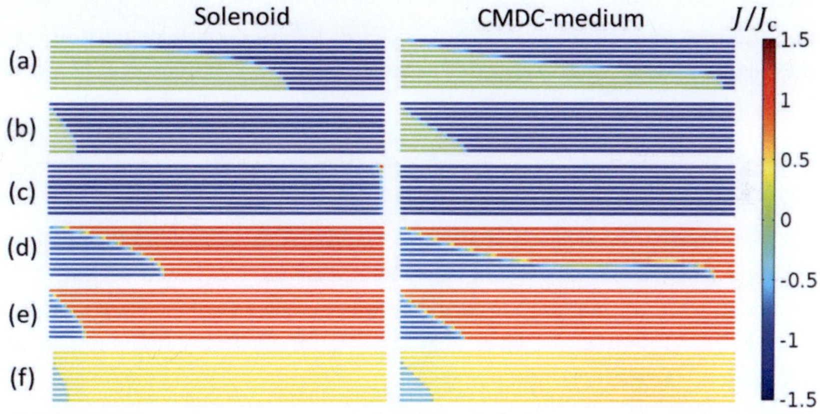

Figure 5.8: Comparison of penetration processes during PFM using solenoid and controlled magnetic density distribution coils with medium gradient of applied field (CMDC-medium). The color scale presents the normalized current density $\boldsymbol{J/J_c}$. (a) to (f) correspond to the time points as illustrated in Fig. 5.7. The plots show one quarter of the sample as marked in Fig. 5.1. The thickness of the YBCO layer is exaggerated for visualization.

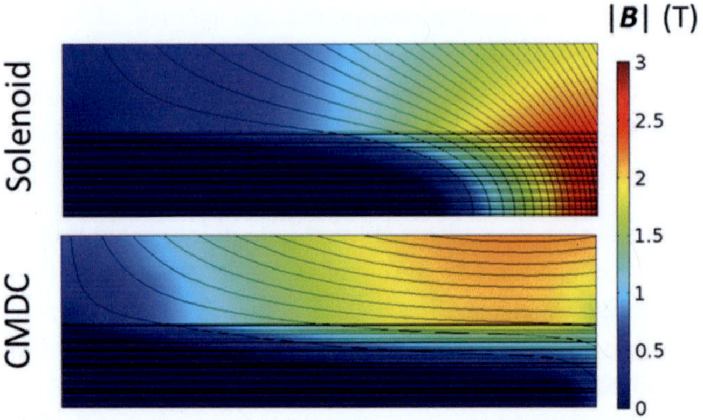

Figure 5.9: Comparison of the magnetic field distribution and flux lines in the stack magnetized by the solenoid and CMDC-medium at time (a) in Fig. 5.7. The plots show one quarter of the sample as marked in Fig. 5.1.

The heat generation powers (unit W/m in 2D) of the different material components during the PFM are plotted in Fig. 5.10. The total heat generation (unit J/m in 2D) during PFM, which is the time integration of each curve in Fig. 5.10, is given in Tab. 5.2. Generally, the heat generation powers are larger in the ascending stage of the pulse than in the descending stage, because the former has a larger variation rate as shown in Fig. 5.7.

For both coils, the superconducting layers generate most of the heat. Copper layers also generate substantial heat due to eddy currents. Silver layers generate less heat because of their small thickness. The heat generation of Hastelloy is considerably lower due to its large electrical resistivity. As a result, tapes with metallic stabilizers of larger resistivity may be more suitable for application in trapped-field magnets, because they generate less heat during PFM. Moreover, the large heat capacity of Hastelloy helps to sink generated heat and to reduce the temperature increase.

Comparing the solenoid and CMDC-medium, the heat generation power by using CMDC-medium is generally lower, not only for the superconducting components, but also for the normal metals. The peaks of heat generation power do not happen at the same time, because of their different penetration processes. At time point (e) in Fig. 5.7, the maximum temperature of the stack is 50 K for CMDC-medium, while 56 K for solenoid. Controlled magnetic density distribution coils reduce the heat generation and reduce the temperature increase compared to solenoid, which also explains why the final trapped field obtained by controlled magnetic density distribution coils is higher than that by the solenoid.

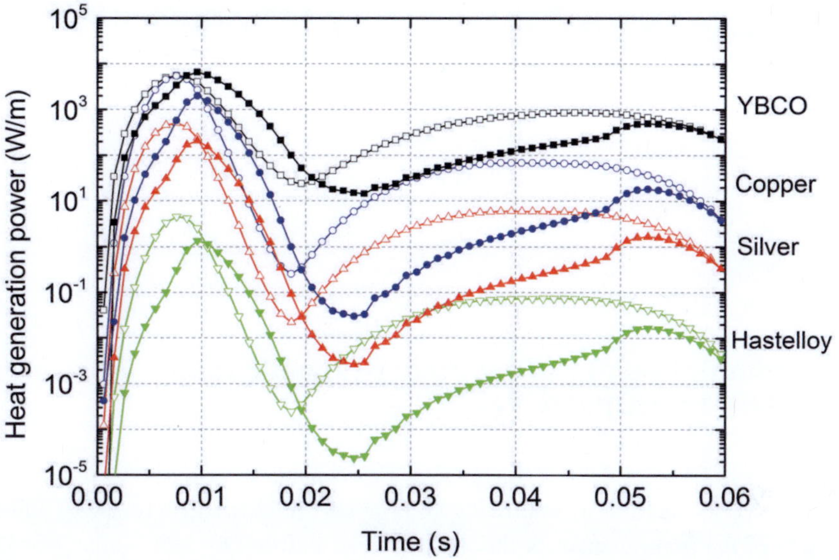

Figure 5.10: Heat generation power of different components during PFM. Empty points and full points represent the solenoid and CMDC-medium, respectively. Four different colors represent four different materials.

In summary, controlled magnetic density distribution coils, which generate gradient applied fields with peak values in the center, can reduce heat generation during PFM and increase the trapped field and trapped flux in the cases simulated in this work, and the larger the gradient, the larger the increase of the trapped field.

The results indicate the possibility of increasing the trapped field by PFM using different coil configurations; however, much more complex engineering is required to build such coils for non-substantial improvement in the trapped field (approximately 10 %). Based on the analysis of dynamics, other possible strategies may be considered, for example, to use ferromagnetic materials to weaken the applied field on the periphery of the stack, or to optimize the overall magnetic field distribution when the stack is situated in a practical application by weakening the applied field on the periphery of the stack.

Table 5.2: Total heat generation of different materials during PFM using a common solenoid or the CMDC-medium coils

	Total heat generation (J/m)	
	Solenoid	CMDC-medium
YBCO	52.0	44.8
Copper	24.1	7.9
Silver	2.4	0.9
Hastelloy	0.0	0.0

5.1.4 Simulation of Pulsed Field Magnetization with Multi-Pulse Technique

As introduced in Chapter 2.6.2.2, it has been experimentally proved that the multi-pulse technique is very effective in improving the trapped field of HTS bulks by PFM. The mechanism of the improvement in the trapped field and flux by multi-pulse technique is not clear yet. The choice of the pulse sequences is mostly based on experience. Generally, the increase of the trapped field by successive pulses is understood as a result of reduction of the temperature rise due to the existence of currents produced by previous pulses. Only a few numerical simulations of multi-pulse technique have been reported, but they are only for HTS bulks. Moreover, the reported work of multi-pulse techniques only simulated several specific pulse sequences, for example [82], [83], but failed to compare all possible sequences.

This chapter provides a systematic simulation study on HTS stacks magnetized by PFM with multi-pulse technique at a fixed temperature including successive pulsed-field application (SPA) and iteratively magnetizing pulsed field method with reducing amplitude (IMRA). The aim of this work is to understand the mechanism of the multi-pulse technique and to suggest an optimal applied pulse sequence. The stack magnetized by three successive pulses of different combinations of pulse magnitudes is simulated to present a complete evolution map of the trapped field and flux by different magnetization sequences. Based on the derived patterns, operable optimal applied pulse sequences are suggested to provide more trapped field and flux.

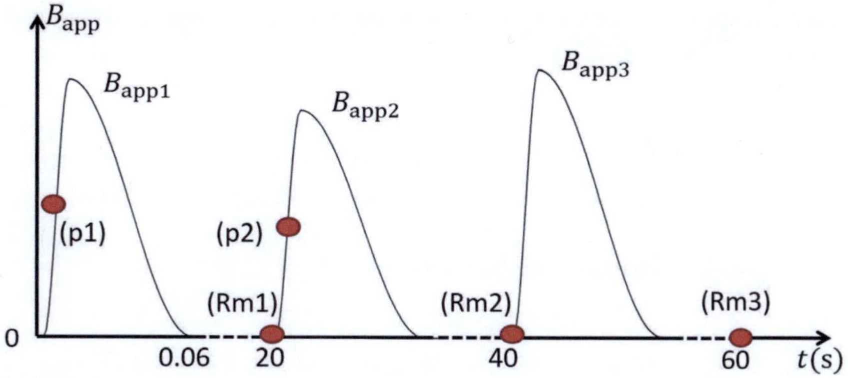

Figure 5.11: Temporal evolution of the three successive applied pulses. The marked time points will be used for later discussion.

The simulations aim to investigate the optimal multi-pulse sequences for PFM, which allows saving the considerable time and costs required by experiments to try out numerous pulse sequences. For verification, experiments carried out on a stack of coated conductors magnetized by different pulse sequences will be presented in Chapter 5.3.2.1. The results in Chapter 5.1.4 and corresponding experiments in Chapter 5.3.2.1 are published in [158].

The model is a 2D electromagnetic-thermal coupled model of an infinitely long stack as described already in Chapters 5.1.1 and 5.1.2. The model simulates a 20-layer stack as shown in Fig. 5.1. Only a quarter of the geometry is simulated to save computation time. The model considers realistic structures of HTS coated conductors with all composing materials (except the buffer layers). The anisotropic magnetic field and temperature dependent critical current density of coated conductors and other temperature-dependent parameters of different materials are comprehensively considered based on the experimental data, as introduced in Chapter 5.1.2. Since the non-linear *E-J* properties of bulks and tapes are generally similar, the derived patterns and conclusions may apply to HTS bulks as well.

5.1.4.1 Evolution of the Trapped Field and Trapped Flux

The stack magnetized by three successive pulses (Pulse 1, 2 and 3) is simulated as shown in Fig. 5.11. There are numerous combinations of the amplitudes of the three pulses. The discussion is arranged to clarify the patterns of how the trapped field and flux evolve after each pulse in order to find out the optimal applied pulse sequences.

The trapped fields B_y are extracted 0.8 mm above the surface center of the stack. The trapped fluxes are defined as the integration of the trapped field 0.8 mm above the stack surface from x=0 mm to x=5 mm, which includes mostly regions of positive remnant fields for the desired conical field distributions. The exact values of trapped flux depend on the chosen region for integration and on the distance between the measurement positions and the stack surface, which can vary in specific applications. The trapped field and flux are extracted 20 s after each pulse (at time Rm1, 2 and 3 in Fig. 5.11) after relaxation, when the temperature has recovered to 30 K.

For reference, the trapped field and flux produced by field cooling or zero field cooling in the stack are 1.377 T and 3.67 mWb/m at 30 K, respectively. These values are calculated using static models by assuming that the current density J inside the stack equals the critical current density J_c everywhere. Detailed descriptions of the models can be found in [140].

PULSE 1

The trapped field and flux at time Rm1 in Fig. 5.11 with the amplitudes of Pulse 1 are shown in Fig. 5.12 and 5.13, respectively. Only representative data points are shown to avoid confusion by excessive data plotting, especially after they evolve with the number of pulses in later discussions. For the trapped field (Fig. 5.12), results are given for B_{app1} = 2.2 T, 2.6 T and 3 T. When B_{app1} equals 1.8 T, only a 0.312 T trapped field is produced, so it is excluded from the figure to avoid unnecessary plotting; for the trapped flux (Fig. 5.13), results are given for B_{app1} = 1.8 T, 2.2 T, 2.6 T and 3 T.

As shown in Fig. 5.12, there is an optimal applied field (2.6 T in Fig. 5.12), which gives the maximum trapped field after Pulse 1. If the applied pulse is too small, the sample is not fully magnetized (namely under-magnetized); however, if the applied pulse is too large, excessive heat is generated and the critical current density is reduced further (namely over-magnetized).

This can be understood better by looking at the normalized current density distribution J/J_c in Fig. 5.14. When under-magnetized, the current density is higher, but there are still large inner regions with negative currents. In this case, the applied field is not large enough, so the negative remnant currents induced during the ascending stage of the applied field still exist. When over-magnetized, the whole stack is filled with positive currents as desired, but the current density is lower due to excessive heat generation. The optimal magnetized situation is a compromise result, when there are still negative currents but the overall current density is relatively high.

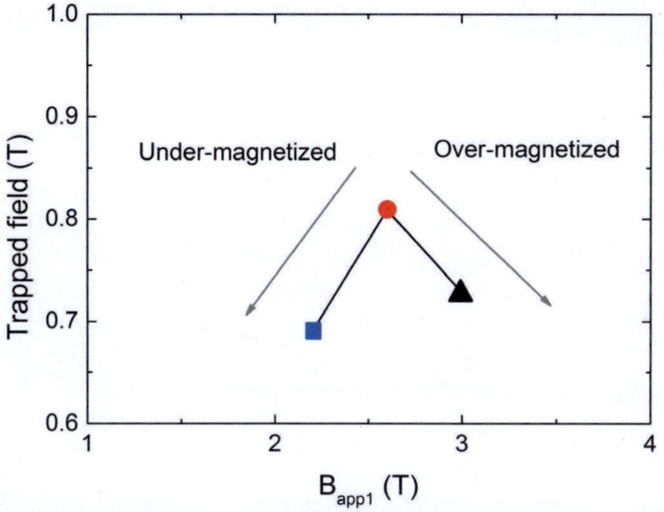

Figure 5.12: The trapped field with the amplitude of Pulse 1 (at time Rm1 in Fig. 5.11).

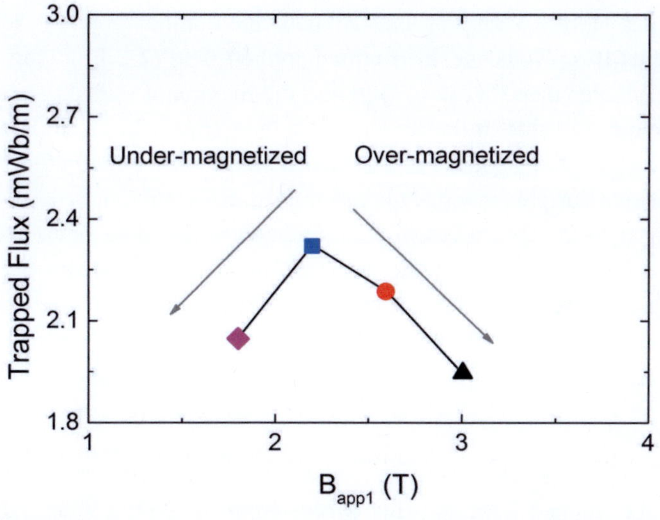

Figure 5.13: The trapped flux with the amplitude of Pulse 1 (at time Rm1 in Fig. 5.11).

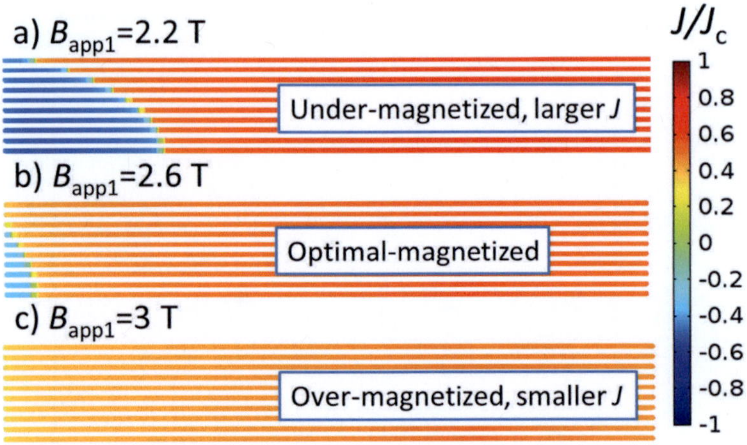

Figure 5.14: The normalized current density distribution after Pulse 1 (at time Rm1 in Fig. 5.11). Only a quarter of the cross-section of the stack is shown, as marked in Fig. 5.1. The thickness of the HTS layer is exaggerated for visualization.

The trapped flux shows a similar pattern with an optimal applied field giving the maximum trapped flux. However, the optimal applied field (2.2 T) giving the maximum trapped flux is smaller than that (2.6 T) giving the maximum trapped field. The trapped field is the single field value in the center, while the trapped flux is an integration of the fields along the interested region. For the tapped field, more penetration (larger applied field) of the positive currents towards the center is more beneficial, because more positive currents are closer to the measurement point and contribute more fields according to the Biot-Savart law. For the trapped flux, larger positive current density on the periphery region contributes more to the total flux, so less penetration (smaller applied field) is preferred. The following discussion will focus on looking for the optimal applied pulse sequence that can produce both the maximum trapped field and flux.

PULSE 2

After Pulse 2, the trapped field and flux evolve from the values after Pulse 1 into new values. The trapped field and flux with the amplitude of Pulse 2 (at Rm2 in Fig. 5.11) are plotted in Fig. 5.15 and 5.16, respectively. Each point in Fig. 5.12 and 5.13 grows into a line in Fig. 5.15 and 5.16, respectively.

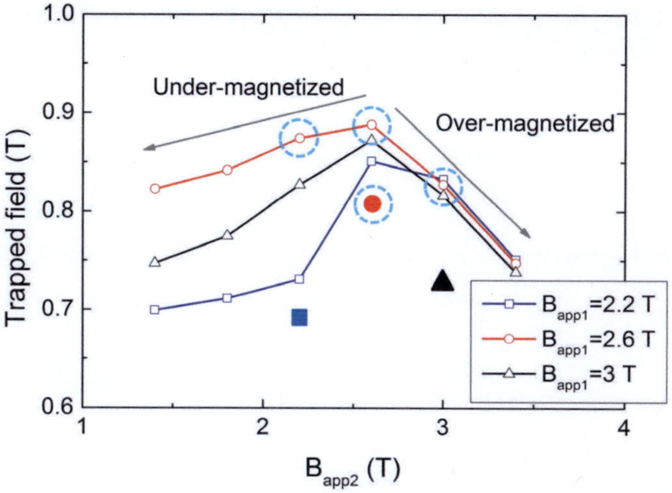

Figure 5.15: The trapped field with the amplitude of Pulse 2 (at time Rm2 in Fig. 5.11). The three full symbols are the trapped fields after Pulse 1 (Fig. 5.12), which grow into the three lines. The marked points will be used in Fig. 5.17.

As shown in Fig. 5.15, there is an optimal applied field for Pulse 2 which gives the maximum trapped field. Otherwise the stack is either under-magnetized or over-magnetized. The optimal applied fields (2.6 T) are the same for the three lines and are also the same with the previous optimal applied field (2.6 T) of Pulse 1. For the trapped flux in Fig. 5.16, there is also an optimal applied field which gives the maximum trapped flux for each line. The optimal applied fields are different for the four lines (1.8 T or 2.2 T). So far, the maximum trapped field is reached by the pulse sequence $B_{app1} = 2.6$ T, $B_{app2} = 2.6$ T; the maximum trapped flux is reached with $B_{app1} = 2.6$ T, $B_{app2} = 2.2$ T.

The normalized current density distributions after Pulse 2 (Rm2 in Fig. 5.11) for typical cases (marked points in Fig. 5.15 and 5.16) are shown in Fig. 5.17. After the second pulse, the current distributions resulted from Pulse 1 evolve into the three situations:

a) Smaller second pulse ($B_{app2} = 2.2 \text{ T} < B_{app1} = 2.6$ T). The penetration is less. Only the currents on the periphery of the stack are influenced. The current density on the periphery is enhanced, because the applied pulse is smaller and the temperature is lower compared to that during Pulse 1. The trapped field is always improved. Yet if Pulse 2 is too small, the improvement is less.

b) Equal second pulse ($B_{app2} = 2.6 \text{ T} = B_{app1} = 2.6$ T). The penetration is less compared to the previous. The current density on the influenced part is improved. The magnetic field and flux line distribution at the same time (5 ms) of the ascending stage of the first and second pulse (time p1 and p2 in Fig. 5.11) are compared in Fig. 5.18. During the magnetization by Pulse 2, the existing currents (resulted from Pulse 1) in the stack expel the flux lines, so the penetration is less compared to Pulse 1, the magnetic flux density is generally lower and the temperature rise is lower. As a result, repeating the same pulse always improves the trapped field.

c) Larger second pulse ($B_{app2} = 3$ T > $B_{app1} = 2.6$ T). The previous penetration is covered. The trapped field can be improved or reduced.

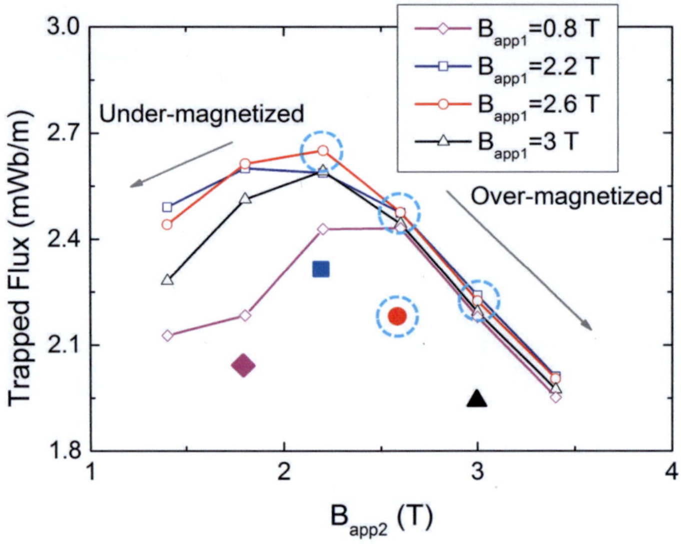

Figure 5.16: The trapped flux with the amplitude of Pulse 2 (at time Rm2 in Fig. 5.11). The four full symbols are the values after Pulse 1 (Fig. 5.13), which grow into the four lines after Pulse 2. The marked points will be used in Fig. 5.17.

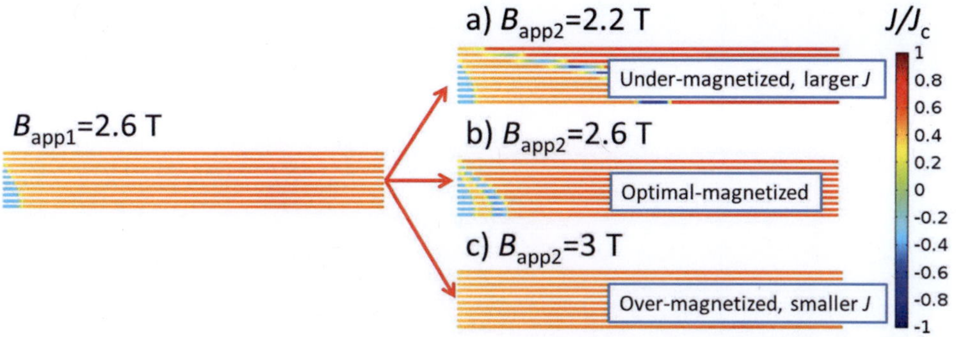

Figure 5.17: The normalized current density distribution (marked points in Fig. 5.15 and 5.16) after Pulse 2 (at time Rm2 in Fig. 5.11). Only a quarter of the cross-section of the stack is shown, as marked in Fig. 5.1. The thickness of the HTS layer is exaggerated for visualization.

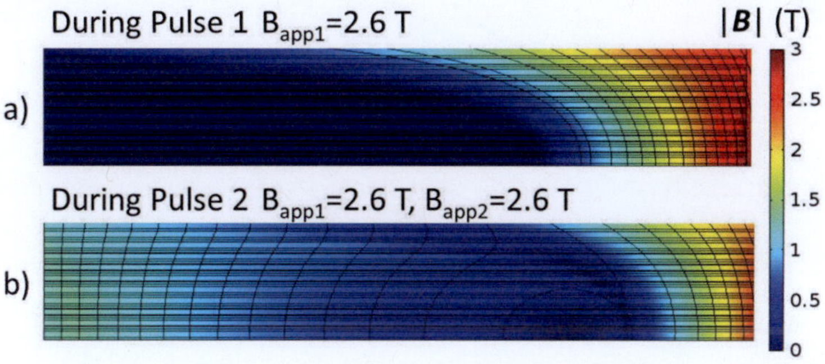

Figure 5.18: The magnetic field and flux line distribution during the ascending stage of Pulse 1 and Pulse 2, which correspond to (a) p1 and (b) p2 in Fig. 5.11. Only a quarter of the cross-section of the stack is shown as marked in Fig. 5.1.

PULSE 3

Similar to the evolution from Pulse 1 to Pulse 2, after Pulse 3 the trapped field and flux evolve from the values after Pulse 2 into new values. To provide an easy way to follow the analysis, an evolution tree describing the trapped field with the pulse number is shown in Fig. 5.19. The trapped field evolves with the magnetization of successive pulses. Each line from Pulse 1 to 3 represents how the trapped field evolves with each applied pulse of a specific pulse sequence. The different colors distinguish three different amplitudes of Pulse 1. The figure does not aim to draw conclusions, but to help provide an overall picture of the performed simulations.

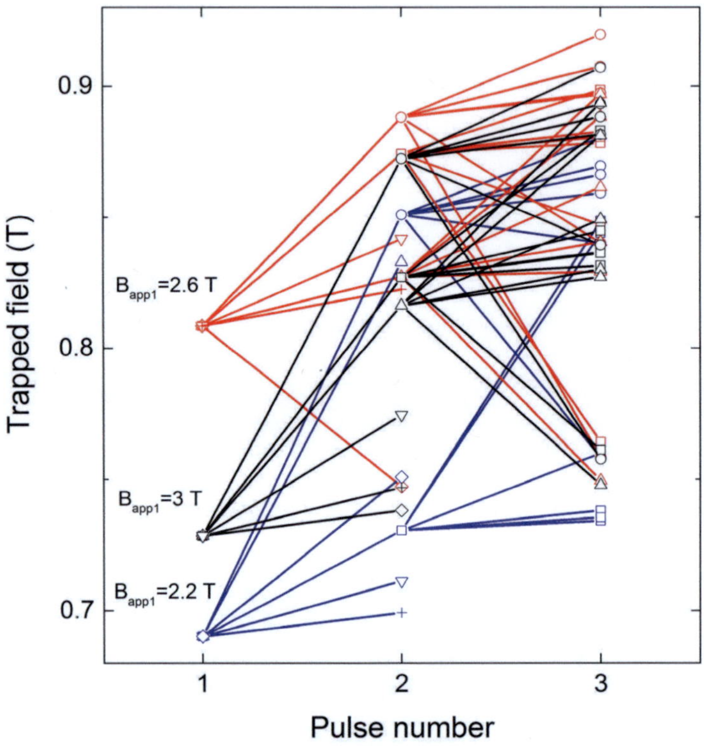

Figure 5.19: The evolution tree: the trapped field with the pulse number. The trapped field evolves with pulse number following different routes, depending on the amplitude of each pulse.

Again, each point in Fig. 5.15 and 5.16 grows into a line in Fig. 5.20 and 5.21, which show the trapped field and flux with the amplitude of Pulse 3 (Rm3 in Fig. 5.11), respectively. To avoid confusion caused by too many lines, only selected data are plotted. For the trapped field, nine points (B_{app1} = 2.2 T, 2.6 T and 3 T; B_{app2} = 2.2 T, 2.6 T and 3 T) in Fig. 5.15 are chosen to generate nine lines in Fig. 5.20. To provide better visualization, the nine lines are divided into three sub figures depending on the applied field of the second pulse B_{app2}. And different colors or symbols distinguish different amplitudes of the first pulse B_{app1}. The large full symbols represent the trapped fields after Pulse 1. The dash lines with empty symbols represent the trapped fields after Pulse 2. These are the same in all subfigures for reference (already used in Fig. 5.12 and 5.15). In each subfigure, the solid lines with blue squares, red circles and black triangles represent that B_{app1} equals 2.2 T, 2.6 T and 3 T, respectively. The trapped fields after Pulse 2 of the solid lines are marked as big empty symbols correspondingly.

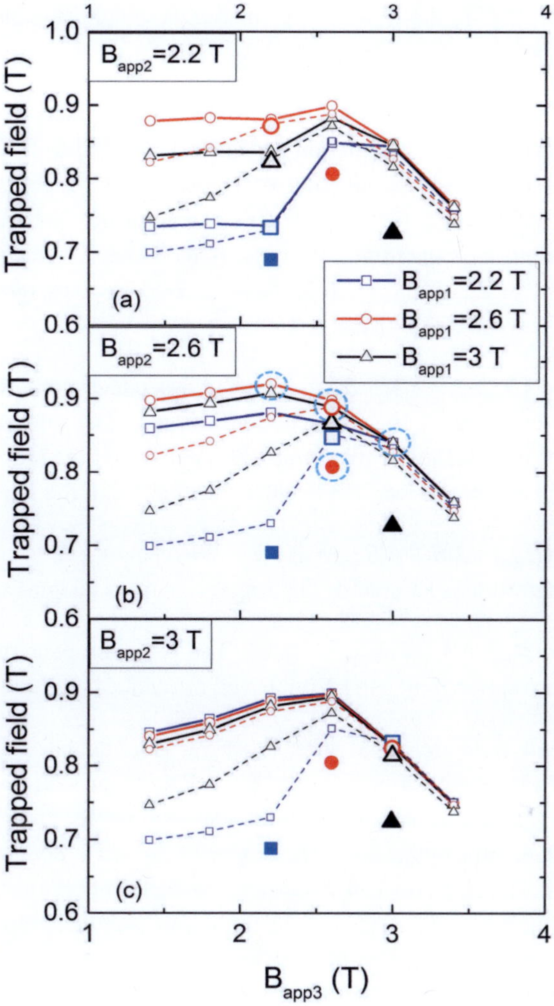

Figure 5.20: The trapped field with the amplitude of Pulse 3 (at time Rm3 in Fig. 5.11). The three full symbols are the values after Pulse 1 (Fig. 5.12), and the dashed lines are the values after Pulse 2 (Fig. 5.15). See text for detailed explanations. The marked points will be used in Fig. 5.22.

Taking the black solid line in Fig. 5.20(a) as an example, the trapped field is the black triangular full symbol at Rm1 ($B_{app1} = 3$ T), then it becomes the strengthened black triangular empty symbol at Rm2 ($B_{app2} = 2.2$ T) and finally becomes the black solid line. Similarly, Fig. 5.21 shows the trapped flux evolved from 16 points ($B_{app1} = 1.8$, 2.2 T, 2.6

T and 3 T; B_{app2} = 1.8, 2.2 T, 2.6 T and 3 T) in Fig. 5.16 with the same visualization technique explained above.

As shown in Fig. 5.20, the trapped field first increases and then decreases with the amplitude of Pulse 3, similar to the situation for Pulse 1 and Pulse 2. The trapped fields are generally improved after the application of Pulse 3. Taking the marked points in Fig. 5.20 and 5.21 as an example, the normalized current density distributions are plotted in Fig. 5.22 for analyzing the penetration process. After Pulse 3, the current distribution resulted from previous pulses (B_{app1} = 2.6 T; B_{app2} = 2.6 T) evolves into three situations again (similar to the discussion in Pulse 2 section):

a) Smaller third pulse (B_{app3} = 2.2 T < B_{app2} = 2.6 T). The penetration is even less. Only the currents on the periphery of the stack are influenced. The current density on the periphery is enhanced, because the applied pulse is smaller and the temperature is lower. The trapped field is always improved. Yet if Pulse 3 is too small, the improvement is less.

b) Equal third pulse (B_{app3} = 2.6 T = B_{app2} = 2.6 T). By repeating the same amplitude for the third time, the penetration is saturated. The improvement is insignificant.

c) Larger third pulse (B_{app3} = 3 T > B_{app2} = 2.6 T). The previous penetration (by Pulse 1 and 2) is covered. The trapped field can be improved or reduced.

The maximum trapped field achieved after Pulse 3 is obtained by using the pulse sequence B_{app1} = 2.6 T, B_{app2} = 2.6 T and B_{app3} = 2.2 T. Combined with the analysis above, the trapped field is improved by magnetizing the stack optimally from the center to the periphery of the stack section by section. The center of the stack demands large pulse to be magnetized, but the large pulse can also increase the temperature and sacrifice the periphery part. After the magnetization of the central part, smaller pulses should be used to magnetize the outer section again without influencing the central part. And this can be done successively until all parts are optimally magnetized.

With the above discussion about penetration, the evolution patterns in Fig. 5.20 can be understood. Figure 5.20(a) has the optimal applied fields at B_{app3} = 2.6 T and shows characteristic points at B_{app3} = 2.2 T. Below 2.2 T the trapped field is not changing much because B_{app3} is too small which only influences the periphery. Above 2.2 T, the trapped field can be increased at 2.6 T by covering part of the previous penetration (B_{app2} = 2.2 T) and make more use of the relatively central part. An even larger Pulse 3 (3 T) will cover the previous penetration and may reduce the trapped field due to more heat generation. At larger Pulse 3 amplitudes, the three lines are overlapping, because all previous penetrations by Pulse 1 and 2 are covered and erased. Figure 5.20(b) has the optimal applied fields at B_{app3} = 2.2 T, because the inner part of the stack has already been optimally magnetized by Pulse 1 and Pulse 2. More trapped field can be acquired by making use of the relatively outer part by applying a smaller pulse. So the optimal applied field is shifted to smaller values. Figure 5.20(c) has the optimal applied fields at B_{app3}=2.6 T. Pulse 2 (3 T) was too large and it has erased Pulse 1, so Pulse 3 can improve the trapped field by magnetizing the central part of the stack optimally with a smaller pulse (2.6 T).

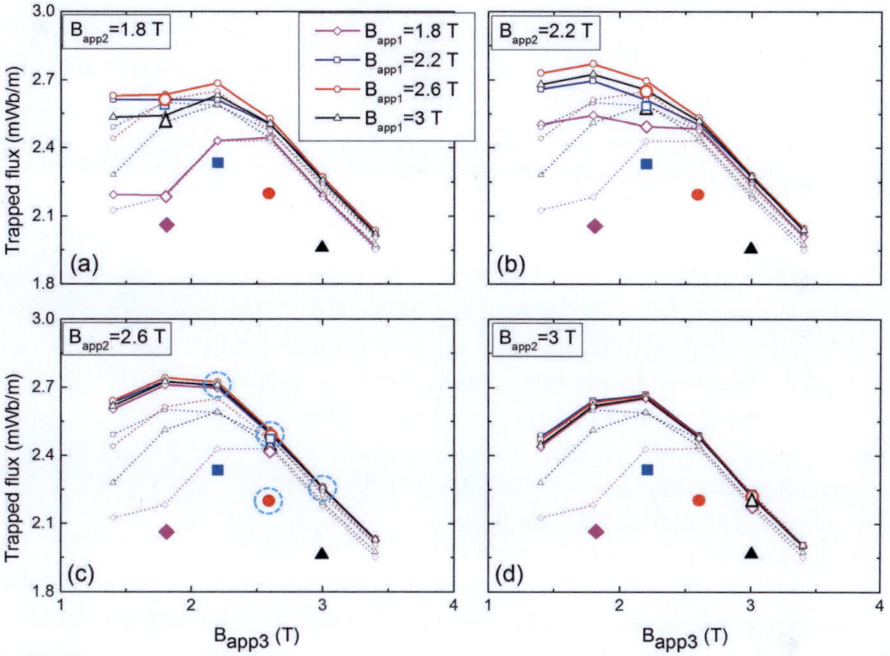

Figure 5.21: The trapped flux with the amplitude of Pulse 3 (at time Rm3 in Fig. 5.11). The three full symbols are the values after Pulse 1 (Fig. 5.13), and the dashed lines are the values after Pulse 2 (Fig. 5.16). See text for detailed explanations. The marked points will be used in Fig. 5.22.

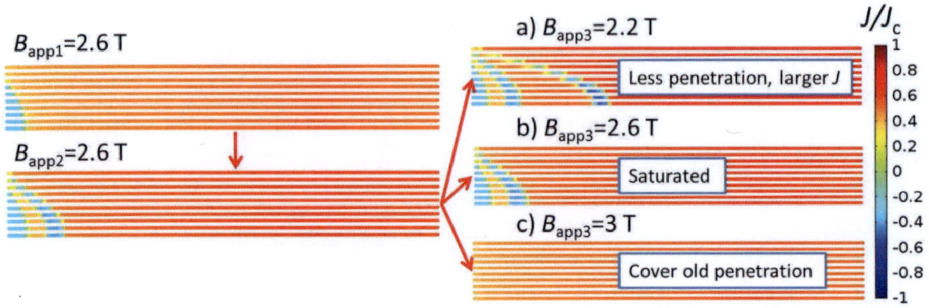

Figure 5.22: The normalized current density distributions (marked points in Fig. 5.20 and 5.21) after Pulse 3 (at time Rm3 in Fig. 5.11). Only a quarter of the cross-section of the stack is shown, as marked in Fig. 5.1. The thickness of the HTS layer is exaggerated for visualization.

Generally speaking, there is an optimal applied field for each applied pulse which gives the maximum trapped field. The optimal applied field will shift to a smaller value after the saturation caused by the previous optimal applied field, which has optimally magnetized the relatively central part of the stack. The saturation is achieved by applying the previous optimal applied field at least once. So Fig. 5.20(a) and (c) still have the optimal applied field 2.6 T, which is the same as that of Pulse 1 and 2; while Fig. 5.20(b) was saturated by the previous optimal applied field (2.6 T) of Pulse 1 and 2 and has a smaller optimal applied field 2.2 T for Pulse 3.

For the trapped flux shown in Fig. 5.21, the patterns are similar with the trapped field, but the optimal applied fields are different. The reason is that the periphery part of the stack contributes more to the total flux compared to the central part. The optimal applied field is fast shifting to smaller values to make more use of the periphery parts of the stack. Figure 5.21(a) has the same pattern as Fig. 5.20(a); Figure 5.21(b) as Fig. 5.20(b); Figure 5.21(c) and (d) as Fig. 5.20(c). Similarly, to achieve more trapped flux, the stack should be magnetized section by section from the center to the periphery to make optimal use of each part of the stack.

5.1.4.2 Optimal Pulse Sequences

Summarizing the analysis in Chapter 5.1.4.1, the conclusions are:

1) For each applied pulse, there is an optimal applied field which gives the maximum trapped field and trapped flux. Otherwise the stack is either under-magnetized or over-magnetized. The optimal applied field for the trapped flux is usually equal or smaller than that for the trapped field.

2) Equal successive pulses can always improve the trapped field or flux, but the improvement comes to saturation.

3) Smaller successive pulses can always improve the trapped field or flux by making more use of the relatively outer part of the stack. If the successive pulse is too small, the improvement will be less.

4) Larger successive pulses can cover the previous penetration. If it is too large, the trapped field or flux can be decreased due to excessive heat generation.

5) The optimal magnetization strategy is to optimally magnetize the stack section by section from the center to the periphery of the stack with pulses of equal or descending amplitudes, so that each section of the stack can be used optimally.

According to these findings, an optimal applied pulse sequence is qualitatively plotted in Fig. 5.23. When a sample is given, the optimal amplitude of the starting pulse is probably unknown. The practical approach is to apply a relatively large starting pulse (B_{m1}) to over-magnetize the stack. Then repeat the same amplitude (Pulse 2, 3 in Fig. 5.23) until saturation is observed. Saturation means that the trapped field and flux are not increasing significantly any more. After the saturation, a smaller pulse (Pulse 4, B_{m2}) should be applied. Theoretically, the reduction in the amplitude of the pulse is infinitesimal. Then repeat the same amplitude again (Pulse 5, 6) until saturation is observed again. Then, Pulse 7 of even smaller amplitude B_{m3} is applied. And the same will be done on and on. In this way, the stack is over-magnetized at the beginning, until the optimal amplitude for the central part is applied. Then, the applied amplitudes keep repeating or decreasing and the stack is optimally magnetized from the center to the periphery section by section.

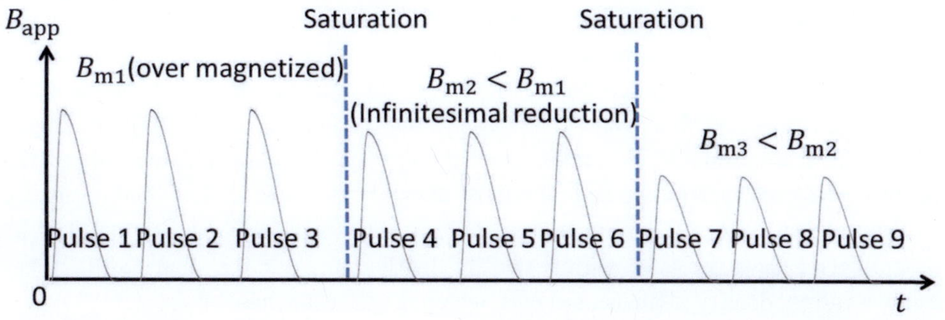

Figure 5.23: Schematic diagram of the optimal applied pulse sequence deduced from the theoretical analysis.

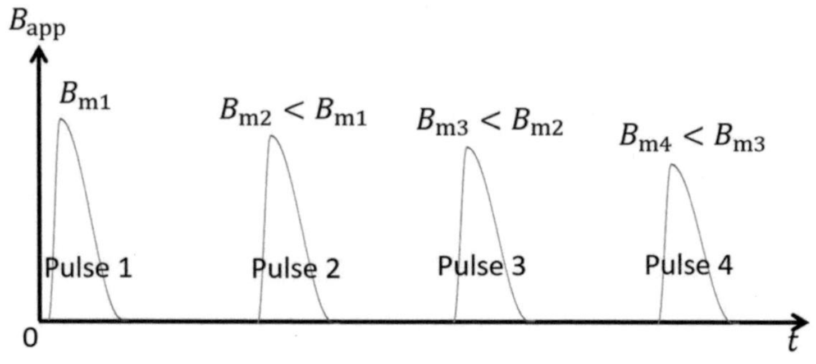

Figure 5.24: Schematic diagram of the optimal applied pulse sequence with a finite number of pulses.

In practice, it is favorable to magnetize a sample with less time or a smaller number of pulses. Considering that the repeated pulses barely improve the trapped field and flux any more after the second pulse, only the pulses with descending amplitudes are necessary, as shown in Fig. 5.24. This will be further justified by the experiments in Chapter 5.3.2.1.

5.1.4.3 Demonstration of Typical Pulse Sequences

Based on the suggested optimal applied sequence in Chapter 5.1.4.2, the stack magnetized by typical pulse sequences are simulated and compared. The trapped field and flux with the amplitude of each pulse is plotted in Fig. 5.25 and 5.26. "3.4/0.2" denotes a pulse sequence which starts from 3.4 T and reduces the amplitude by 0.2 T for each pulse. And so on, for other sequences. The trapped field and flux evolve along each line from the right to the left with each applied pulse.

Figure 5.25 compares four pulse sequences with different starting amplitudes. The amplitudes reduce by 0.2 T for each successive pulse. Generally, the trapped field and flux improve with each applied pulse. The larger the starting amplitude, the larger the final trapped field and flux. When the starting amplitude is over 3 T, the final trapped field and flux are saturated. When the starting amplitude is below 3 T, the trapped field and flux are lower because the central part of the stack is not used. The final normalized critical current density J/J_c of the four sequences are shown in Fig. 5.27. "3.4/0.2" and "3/0.2" have almost the same current density distribution; "2.6/0.2" and "2.2/0.2" have negative current density in the central part, which is under-magnetized.

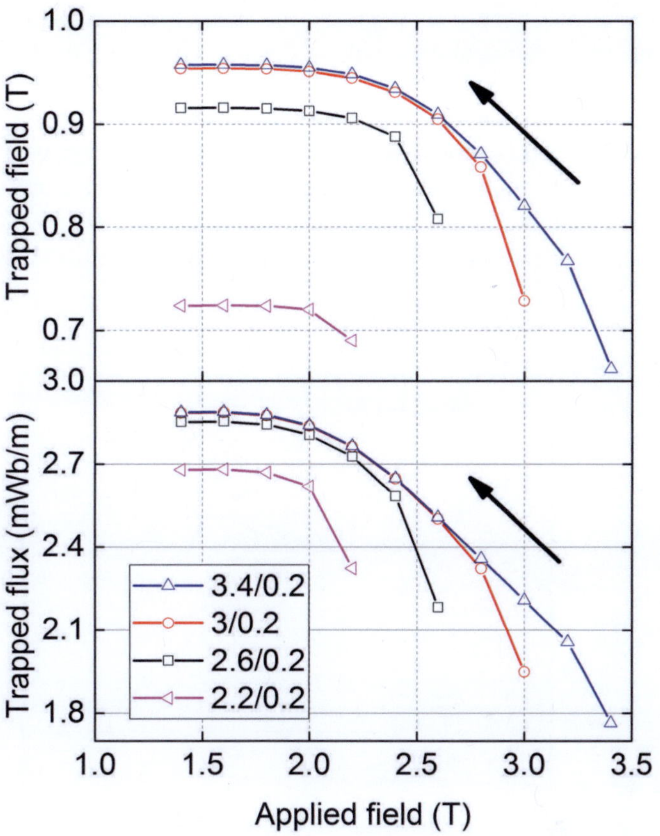

Figure 5.25: Demonstration of pulse sequences with different starting amplitudes. "3.4/0.2" denotes a pulse sequence which starts from 3.4 T and reduces the amplitude by 0.2 T for each pulse. And so on, for other sequences. The trapped field and flux evolve along each line from the right to the left with each applied pulse.

Figure 5.26 compares pulse sequences that start with the same amplitude (3 T) but have different amplitude intervals. In particular, "3/0.4×3" denotes that each amplitude repeats for three times and then reduces by 0.4 T in the sequence. "3/0.2" leads to more trapped field and flux than "3/0.4×3" with a smaller number of pulses, so it is not necessary to repeat the same amplitude considering that less pulses are preferred to save time and costs in practice, as mentioned in Chapter 5.1.4.2.

Comparing the four pulses that start with 3 T and reduce the amplitude by different intervals, one can observe that the smaller the amplitude interval, the larger the final trapped field and flux. This is consistent with the deduction in Chapter 5.1.4.2. As shown in Fig. 5.28, with smaller amplitude intervals, the stack is magnetized by finer sections. The improvement can reach saturation with the decrease of the amplitude interval.

In summary, based on the numerical simulations presented in this chapter, the following optimal applied sequence is proposed: first apply a pulse of large amplitude to over-magnetize the stack; then apply pulses of smaller amplitude (theoretically reduced by an infinitesimal amount at each step) to optimally magnetize the stack section by section. Experiments were carried out to further verify the simulations as presented in Chapter 5.3.2.1.

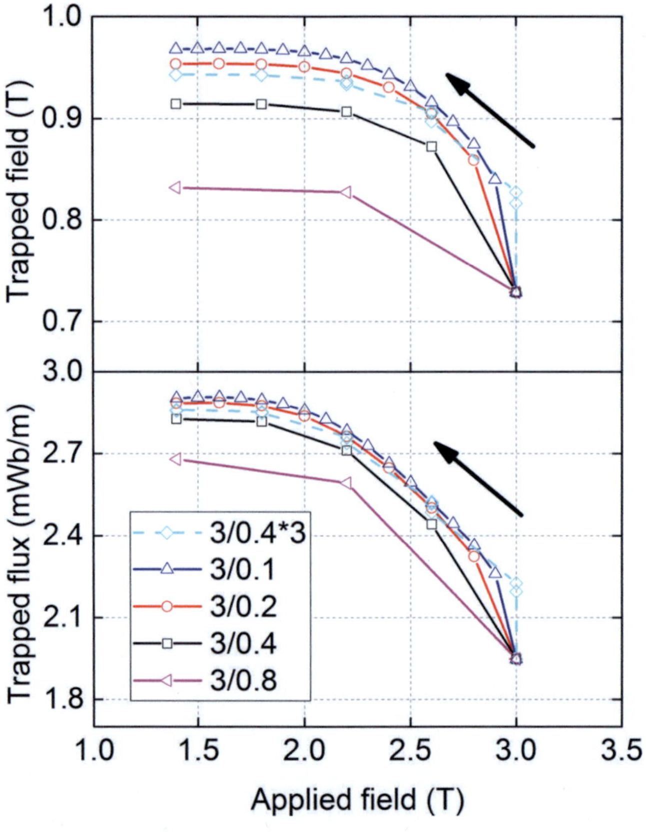

Figure 5.26: Demonstration of pulse sequences with different amplitude intervals. "3/0.1" denotes a pulse sequence which starts from 3 T and reduces the amplitude by 0.1 T for each pulse. And so on, for other sequences. "3/0.4×3" denotes that each amplitude repeats for three times in the sequence. The trapped field and flux evolve along each line from the right to the left with each applied pulse.

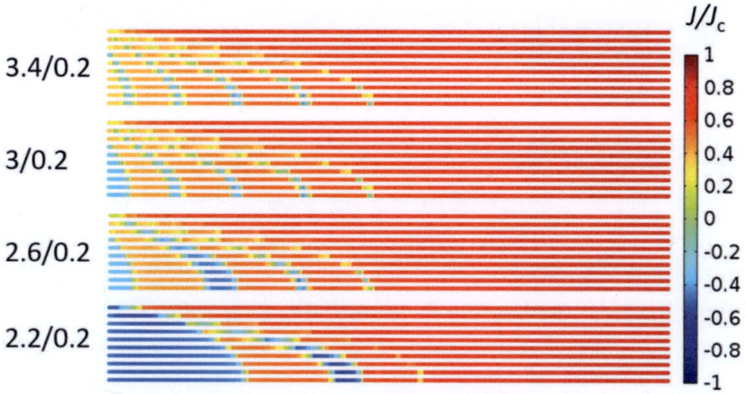

Figure 5.27: The final normalized critical current density J/J_c obtained by the four pulse sequences with different starting amplitudes corresponding to Fig. 5.25. Only a quarter of the cross-section of the stack is shown, as marked in Fig. 5.1. The thickness of the HTS layer is exaggerated for visualization.

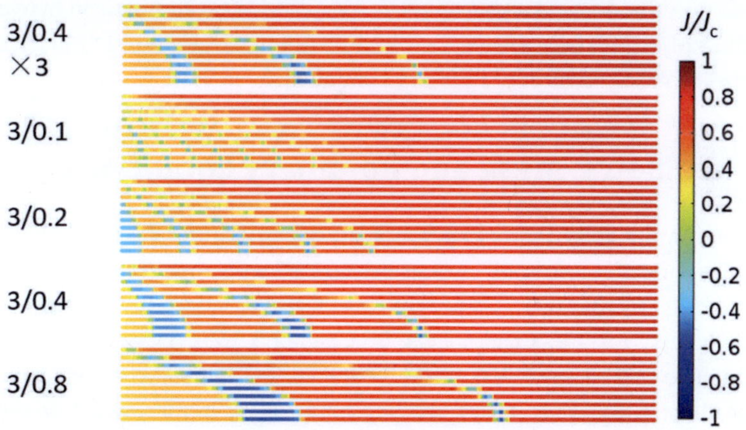

Figure 5.28: The final normalized critical current density J/J_c obtained by the four pulse sequences with different amplitude intervals corresponding to Fig. 5.26. Only a quarter of the cross-section of the stack is shown, as marked in Fig. 5.1. The thickness of the HTS layer is exaggerated for visualization.

5.2 3D Modelling

2D modelling provides good approximations qualitatively and quantitatively for many practical cases with low cost in terms of computation time; however, it is not able to describe phenomena caused by 3D spatial effects. 3D models with the ***H***-Formulation has have been developed to simulate HTS designs such as twisted filaments [159], [160], racetrack coils [161], and Roebel cables [162], when symmetries do not allow a 2D reduction of the problem. Due to a dramatic increase in the number of degrees of freedom, 3D modelling is computationally much more challenging than 2D modelling at the same level of accuracy or discretization. Homogenization techniques have been developed to improve the computing speed with a desirable accuracy [161].

Stacks of HTS coated conductors are often presented as cuboids, which are not axisymmetric, although in the literature 2D axisymmetric models are often used for simplification [84], [144]. The approximation of a curved stack by 2D models is even more questionable. Accurate simulations of HTS stacks require 3D models.

In this chapter, a static 3D model is first presented to simulate flat and curved HTS stacks magnetized by field cooling or zero field cooling magnetization. Then a dynamic 3D electromagnetic-thermal coupled model extended from the previously used 2D model is presented to describe HTS stacks magnetized by PFM. The comparison between 2D and 3D modelling of a flat stack magnetized by FPM is made. Using 3D modelling, the flat and curved stacks magnetized by PFM are compared.

5.2.1 Geometry, Mesh and Boundary Conditions

In this chapter, a flat and a curved stack are considered for the 3D modeling as shown in Fig. 5.29. The curved stack is of interest because of its geometrical applicability for the rotor in an electrical machine. To provide an intuitive picture, the photo of the curved stack that is used for the experiments in Chapter 5.3.2.2 is shown on the right of Fig. 5.29(b). The curved stack was made by attaching a flat stack closely onto a wooden rod. In this chapter, the diameter of bending is set to be 20 mm, which is the minimum bending diameter for non-degradation of the critical current density [26]. The static model for simulation of static field magnetization in Chapter 5.2.3 is not time consuming, so real geometries are considered and meshed with free tetrahedral elements. In the following, the construction of the dynamic model for simulations of PFM in Chapter 5.2.4 is explained.

Considering symmetry, modelling can be carried out for only part of the geometry to save computation time especially for the 3D modelling of PFM. As shown in Fig. 5.30(a), for the 3D modelling of a flat stack magnetized PFM only one eighth of the stack is simulated. In this chapter, a square stack of 15 layer coated conductors magnetized by PFM is modelled to compare with the experimental results reported in [84]. The formulations of the model were described in Chapter 3.3.3. The pieces coated conductors are 62 μm thick, 12 mm wide, 12 mm long and produced by SuperOx [163].

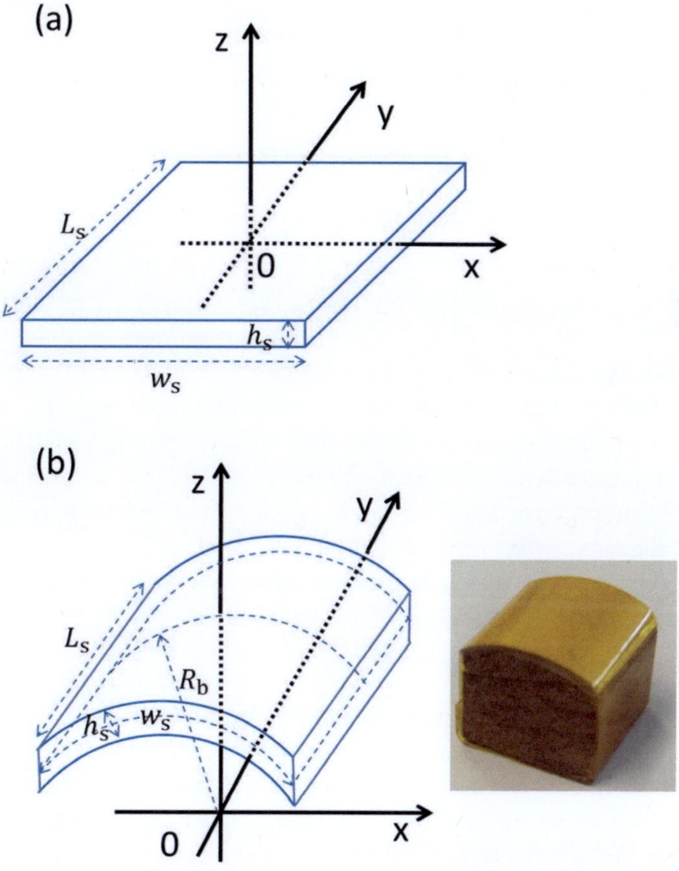

Figure 5.29: Geometries of stacks for comparison. (a) A flat stack and (b) a curved stack are considered for 3D modeling. For both stacks, w_s, L_s ($w_s \leq L_s$) and h_s are the width, length and thickness of the stack, respectively. For the simulations and experiments in the following chapters, a square stack is assumed when $w_s = L_s = 12$ mm. The bending radius of the curved stack is R_b. The photo of a curved stack attached to a wood rod ($R_b = 10$ mm) is shown.

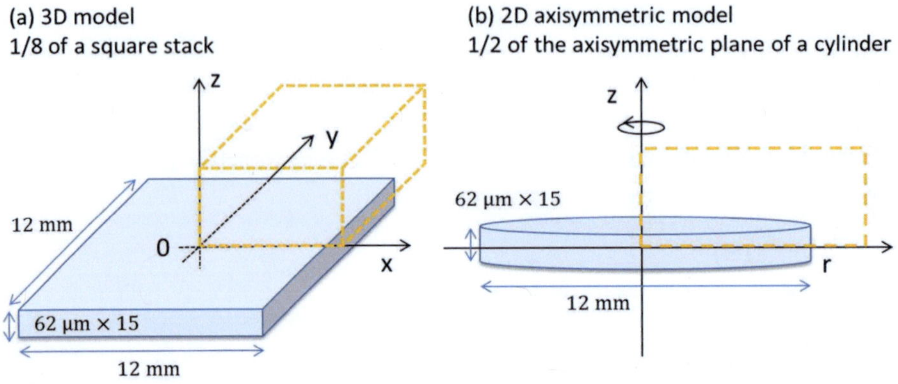

Figure 5.30: The geometries used for 2D and 3D simulations. Simulations with (a) 3D and (b) 2D modelling are carried out for a stack of 15-layer coated conductors. For the 3D modelling, one eighth of the stack is modelled; for the 2D axisymmetric modelling, one half of the axisymmetric plane is modelled.

The stack is homogenized without considering the layer-by-layer structures. In order to reduce the number of mesh elements for the stack, two mesh transition layers, namely "jackets", are added surrounding the stack [161] as shown in Fig. 5.31. The HTS stack is meshed with hexahedral elements and the outmost air domain is meshed with tetrahedral elements. The jacket layer attached to the stack is meshed with pyramid or prism elements and the outer jacket layer is meshed with tetrahedral elements. In this way, the hexahedral elements of the stack and the tetrahedral elements of the outmost air domain are connected.

Since the stack is homogenized in geometry to become a bulk, an extra constraint is required for the simulation. The stack is essentially different from a bulk because the stack is composed of layer-by-layer tapes that contain Hastelloy substrate; as a result, there is no current flowing between layers. The currents can be constrained parallel to the stack surface by placing air gaps between layers, combining integral constraint with proper mesh or defining anisotropic electrical resistivity as suggested in [161]. In this work, the condition is imposed by revising the equation of the definition of electrical field (Eq. 3.17c) for the HTS domain as

$$J_z = 0. \tag{5.4}$$

This equation is imposed in the simulation in the variable definition of the stack domain. For the air domain, the definition of J_z remains as Eq. 3.17c.

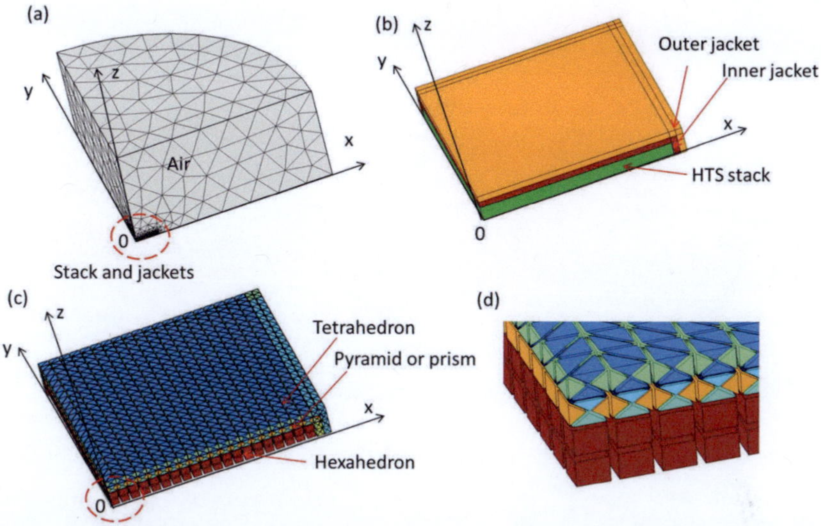

Figure 5.31: The mesh for the flat stack . (a) The 3D mesh of one eighth of the flat square stack. (b) Two transition layers are added surrounding the HTS stack, namely "jackets". (c) The HTS stack is meshed with hexahedral elements; the attached jacket is meshed with pyramid or prism elements; the outer jacket and the outmost air domain are meshed with tetrahedral elements. The elements shrink with a factor of 0.8 to provide visualization. (d) Zoom-in of the mesh at the corner of the stack.

The axisymmetric 2D model is the common approach to model square stacks for simplification [84], [144], although the model actually describes a cylinder rather than a cuboid. A comparison is made for comparing the 2D and 3D modelling of a flat stack magnetized by PFM in Chapter 5.2.4.1. For the axisymmetric 2D modelling, one half of the axisymmetric plane is modelled as shown in Fig. 5.30(b). Correspondingly, magnetic flux is set perpendicular to the boundary on the z-axis line. The model is similar to that for HTS bulks described in Chapter 4.

For the curved stack magnetized by PFM, the simulation is carried out for one fourth of the geometry ($x>0$ and $y > 0$ in Fig. 5.29(b)). The curved stack is assumed to be made from the same stack but curved around the y axis with a bending diameter of 20 mm as shown in Fig. 5.29(b).

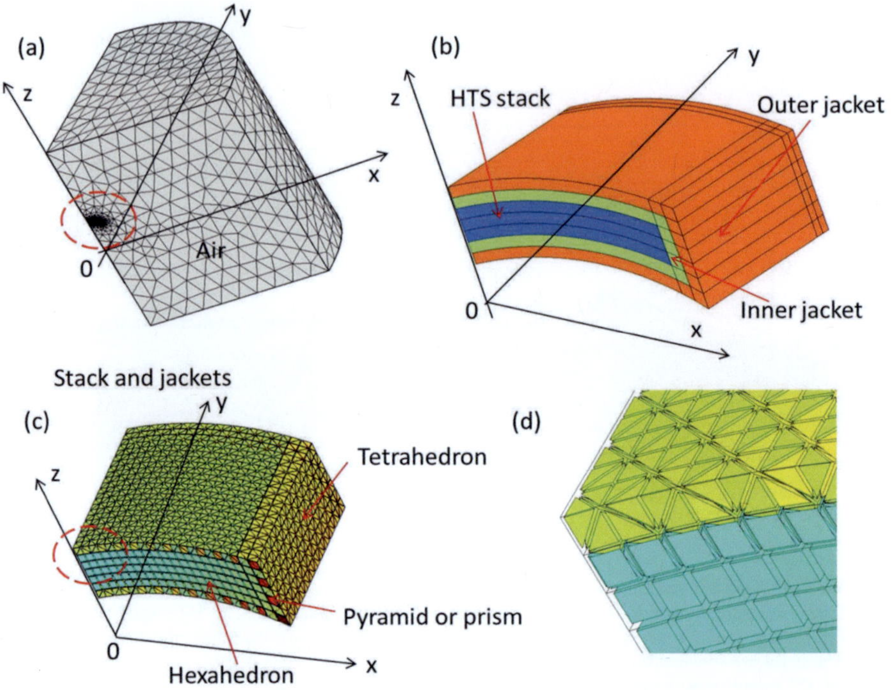

Figure 5.32: The mesh for the curved stack. (a) The 3D mesh of one fourth of the curved stack. (b) Two layers of jackets are added surrounding the HTS stack. (c) The HTS stack is meshed with hexahedral elements; the attached jacket is meshed with pyramid or prism elements; the outer jacket and the outmost air domain are meshed with tetrahedral elements. The elements shrink with a factor of 0.8 to provide visualization. (d) Zoom-in of the mesh at the corner of the stack.

The curved stack is also homogenized as an entire bulk without considering laminated tapes to save computation time. The stack domain is meshed with similar techniques for the flat stack as shown in Fig. 5.32.

Similar to the flat stack, currents should not flow between tapes due to the existence of Hastelloy substrates. The current can only flow in the tape surfaces, which means that the current along the curving radius of the curved stack should be zero

$$J_r = J_z \frac{z}{\sqrt{x^2 + z^2}} + J_x \frac{x}{\sqrt{x^2 + z^2}} = 0 \, , \qquad \textbf{(5.5)}$$

where J_x, J_z and J_r are current components using the coordinate in Fig. 5.29(b). This equation is imposed by revising the equation of the definition of electrical field (Eq. 3.17c) for the HTS domain as

$$J_z = -\frac{xJ_x}{z}, \tag{5.6}$$

For the electromagnetic model, the magnetic flux density on the outmost boundary of the modelling domain is prescribed as the external applied field. In this chapter, the applied field is parallel to the z axis and the duration of the pulsed field is approximately 28 ms, consistently with the experiments in Chapter 5.3.2.2. The shape of the pulse is given by

$$B_{\text{app}} = \frac{1}{2}\left(1 - \cos\left(\frac{\pi t}{\tau}\right)\right) B_{\text{m}}, \qquad 0 \le t \le 2\tau \tag{5.7}$$

where B_{m} is the peak applied field and $\tau = 14$ ms is the characteristic time when the applied field reaches the peak value. The shape of this pulse is different from previous chapters (see Fig. 5.36). Considering symmetry, the magnetic flux is set parallel to the boundary on the x = 0 or y = 0 plane.

For the thermal model, the boundary condition is not easy to determine because of the uncertainty in the measurement and in the set point temperature of the cooling head [73]. According to the authors of [84], the stack was not directly attached to a cold head, but surrounded by helium gas, which provides convective cooling. The temperature of the cold head is not set to be the supposed equilibrium temperature T_0, but to be a certain temperature lower than T_0 to improve the speed of cooling. After the magnetization, the temperature will decrease due to the cooling; when the temperature reaches T_0, the trapped field is measured and the next programmed pulse is triggered. The set point temperature of the cold head is based on experience to make the recovery of the temperature within minutes and also to avoid over-shooting the temperature to be lower than T_0. For simplification and referring to [85], the heat flux density on the boundary of the stack for cooling in the simulation is imposed as

$$\phi_{cool} = \begin{cases} 100\left(\frac{T - T_0}{15}\right) \text{W/m}^2, & \text{if } T \le T_0 + 15 \\ 100\,\text{W/m}^2, & \text{if } T > T_0 + 15 \end{cases} \tag{5.8}$$

where T_0 is the working temperature. The cooling flux is positively correlated to T_0 because the cooling is provided by the convection of the helium gas. A saturation cooling flux is supposed, considering that the temperature of the helium gas around the sample can be heated up and the cooling efficiency can be decreased. The cooling is supposed to happen only when the temperature is higher than T_0, because over shooting of the temperature below T_0 is avoided in the experiments. Trial simulations show that the temperature recovery time and trapped fields are close to measurements, which justifies the simplified cooling condition.

5.2.2 Material Properties

For comparison with the experiments reported in [84], in the simulation both the magnetic field and temperature dependences of the critical current density of the HTS coated conductors are considered with the Kim model

$$L(B,T) = \frac{J_c(B,T)}{J_{c0}(\text{self}, 77\ \text{K})} = \frac{l_{c0}(T)}{1 + \frac{B}{B_0(T)}}, \tag{5.9}$$

where B is the magnitude of magnetic field and $l_{c0}(T)$ and $B_0(T)$ are temperature-dependent parameters. Similar to Chapter 5.1.2, the measured lift factor data are fitted for each temperature to determine $l_{c0}(T)$ and $B_0(T)$. Then the parameters for other temperatures are obtained by direct interpolation of the parameters with the temperature.

The parameters fitted from the measured data [164] provided by SuperOx are listed in Tab. 5.3. The measured lift factors and the fitted ones for given temperatures are shown in Fig. 5.33. In the simulation, $J_{c0}(\text{self}, 77\ \text{K})$ 3.3×10^{10} is used, which corresponds to a self-field critical current of 400 A at 77 K for a 12 mm-wide tape. The power exponent n is 21. In order to facilitate the convergence of the model, the resistivity is made constant when J/J_c is larger than 1.6. The thermal conductivity λ and the heat capacity c_p of the stack are based on those of the Hastelloy provided in Chapter 5.1.2, as the Hastelloy constitutes 97 % of the entire mass of the tape.

Table 5.3: Fitted parameters of the lift factor of a SuperOx tape with Eq. 5.9 for given temperatures

T (K)	l_{c0}	B_0 (T)
5	10.84	1.48
20	8.15	0.96
40	5.01	0.65
65	2.30	0.30
77	1.00	0.18

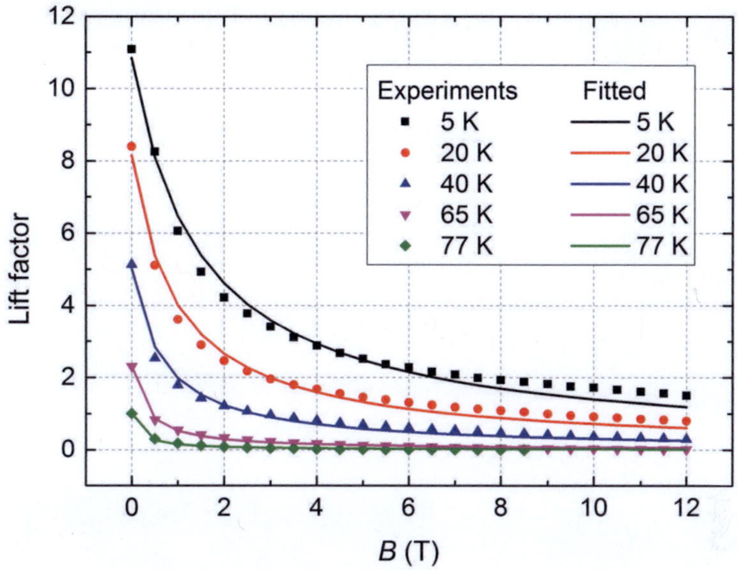

Figure 5.33: The measured lift factors and the fitted lift factors of a SuperOx tape with the field amplitude B using Eq. 5.9 with parameters listed in Tab. 5.3 for given temperatures [164].

5.2.3 Simulation of Static Field Magnetization

A static model is used to describe the final state of stacks after magnetization with field cooling or zero field cooling. The magnetization process of field cooling or zero field cooling can also be simulated with dynamic models as described in Chapter 3.3 and the magnetic field dependence of J_c can be taken into account [16], [144]. Such simulation, especially when it comes to 3D, is rather time consuming. One 3D dynamic simulation may take hours to days to run. The main advantage of using a static model instead of the dynamic model is that the trapped field profiles can be estimated within minutes even with a fine mesh discretization. In the static modelling, the remnant current density distribution in the stack is prescribed according to the critical state model as described in Chapter 2.4.1. Then the static magnetic field distribution is calculated according to Ampere's law using the Magnetic Field (mf) module in COMSOL [112].

According to the critical state model, after magnetization the HTS stack sustains a loop-like remnant current, the density of which is the critical current density. The current distribution follows the geometrical boundaries of the stack, which was proved by magneto-optical imaging [165]. Ignoring the magnetic field dependence of the critical current density, the remnant current distribution for the flat stack in Fig. 5.29(a) is,

$$J_{\mathrm{flat}}(x,y) = J_{\mathrm{c}}^{*} \times \begin{cases} \hat{x}, & \text{if } y < -|x| - \frac{1}{2}(L_{\mathrm{s}} - w_{\mathrm{s}}) \\ \hat{y}, & \text{if } x > |y| - \frac{1}{2}(L_{\mathrm{s}} - w_{\mathrm{s}}) \\ -\hat{x}, & \text{if } y > |x| + \frac{1}{2}(L_{\mathrm{s}} - w_{\mathrm{s}}) \\ -\hat{y}, & \text{if } x < -|y| + \frac{1}{2}(L_{\mathrm{s}} - w_{\mathrm{s}}) \end{cases} \tag{5.10}$$

and the remnant current distribution for the curved stack in Fig. 5.29(b) is,

$$J_{\mathrm{curved}}(x,y) = J_{\mathrm{c}}^{*} \times$$

$$\begin{cases} \frac{z}{\sqrt{x^2+z^2}}\hat{x} - \frac{x}{\sqrt{x^2+z^2}}\hat{z}, & \text{if } y < -|x| - \frac{1}{2}(L_{\mathrm{s}} - 2R_{\mathrm{b}}\sin(\frac{\alpha}{2})) \\ \hat{y}, & \text{if } x > |y| - \frac{1}{2}(L_{\mathrm{s}} - 2R_{\mathrm{b}}\sin(\frac{\alpha}{2})) \\ -\frac{z}{\sqrt{x^2+z^2}}\hat{x} + \frac{x}{\sqrt{x^2+z^2}}\hat{z}, & \text{if } y > |x| + \frac{1}{2}(L_{\mathrm{s}} - 2R_{\mathrm{b}}\sin(\frac{\alpha}{2})) \\ -\hat{y}, & \text{if } x < -|y| + \frac{1}{2}(L_{\mathrm{s}} - 2R_{\mathrm{b}}\sin(\frac{\alpha}{2})) \end{cases} \tag{5.11}$$

where $\hat{x}$, $\hat{y}$ and $\hat{z}$ are the unit vectors along the x, y and z directions, respectively; α equals $w_{\mathrm{s}}/R_{\mathrm{b}}$, which is the angle of curvature of the curved stack; and J_{c}^{*} is the modified critical current density considering flux creep. Due to flux creep [41], [133], [135], the electrical field in a stack after magnetization cannot maintain at E_{c0}. The remnant current and trapped field decay with time (for example, see Fig. 4.10). The modified critical current density J_{c}^{*} is the critical current density with a modified criterion electrical field E_{c}^{*}, which can be derived from the *E-J* power law in Eq. 2.5

$$J_{\mathrm{c}}^{*} = J_{\mathrm{c0}}\left(\frac{E_{\mathrm{c}}^{*}}{E_{\mathrm{c0}}}\right)^{\frac{1}{n}}, \tag{5.12}$$

where J_{c0} is usually the critical current density measured when $E_{\mathrm{c0}} = 10^{-4}$ V/m. When the trapped field is measured in experiments, the electrical field E_{c}^{*} in the sample is usually lower than the standard characteristic field E_{c0} due to flux creep [133], [135]. To compare with experiments, a modified criterion electrical field E_{c}^{*} will then be used.

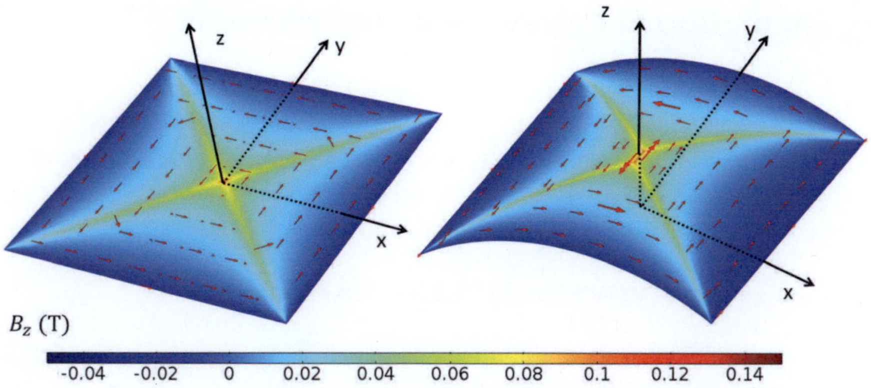

Figure 5.34: The current vector arrows and the z-direction magnetic flux density distribution on the volume surface for (a) a flat stack and (b) a curved stack after magnetization, respectively.

A square flat stack and a curved one are simulated as a demonstration for the model. The stacks are made of 0.035 mm-thick 12 mm-wide coated conductors of nominal critical current 450 A at 77 K. The bending diameter of the curved stack is 20 mm. The model assumes that the current density in the stacks is uniformly distributed and that the stacks are homogenized with no laminated structures. The current density is obtained by directly dividing the nominal critical current by the cross-section area of the tape.

The current vector arrows and the z-direction magnetic flux density (B_z) distribution on the volume surface are shown in Fig. 5.34. The current distributions are loop-like, following the geometric boundaries of the stack as prescribed, and there is no inter-layer current as imposed by Eq. 5.10 and 5.11. The peak value of B_z is in the center of the stacks and B_z is negative on the periphery of the stack. Due to the curvature, the width of the curved stack along the x axis is shorter compared to the flat stack. Accordingly, the magnetic field distribution along the x axis is contracted, which will be further explained in Chapter 5.3.1.

The static 3D modelling can be used for quickly estimating the maximum possible trapped field in a stack produced by field cooling or zero field cooling. This approach does not consider the magnetic field dependence of the critical current density, and is valid when the stack is thin and the self-field is not significant. A further discussion including quantitative comparison between the simulation and experiments is presented in Chapter 5.3.1.

5.2.4 Simulation of Pulsed Field Magnetization

The pulsed field magnetization process of stacks of HTS coated conductors has to be described by dynamic electromagnetic-thermal coupled models. First, an ordinary flat stack magnetized by PFM is simulated to compare 2D and 3D models and the results are compared with experiments reported in the literature. Then a curved stack and a flat stack are simulated and compared using 3D models.

5.2.4.1 Comparison of 2D and 3D Modelling of a Flat Stack

3D models provide a more realistic description of an object of a finite geometry than 2D models; however, the accuracy is often compromised to reduce computation time to an acceptable extent. In many situations, 2D models can provide an overall fair estimation when 3D phenomena are not of concern.

In this chapter, an ordinary square flat stack magnetized by PFM is simulated with both 2D and 3D models for comparison. The construction of the model has been introduced in Chapter 5.2.1 and 5.2.2.

The simulation is compared with reported experimental results in [84]. The original experiments were carried out using the MPSC method (multi-pulse technique with step-wise cooling). The stack was first magnetized at 77 K with pulses of reduced amplitudes; then, it was cooled down to a lower temperature and magnetized again. For simplification, only one pulse was simulated separately for each temperature. The trapped fields were extracted 0.8 mm above the surface of the stack as in the experiments. The experimental results were extracted after the first pulse at each equilibrium temperature from the raw data of stacks with 1 μm silver stabilizer from the raw data of [84]. The time for extraction of the trapped field is when the sample is cooled down to the equilibrium temperature and the next pulse is about to be triggered. The applied field amplitudes and the time for extraction of the trapped fields in the simulation for different temperatures are listed in Tab. 5.4.

The calculated trapped fields with the equilibrium temperature using four models are compared in Fig. 5.35. The four models are 3D and 2D models with or without coupling the heat equations. The applied fields for the 2D and 3D electromagnetic models at 15 K are 5 T for the simulation, because 3.1 T is not enough to fully magnetize the stack when the thermal effect is not considered. The trapped fields obtained by applied field of 3.1 T are kept as dash lines for reference.

Table 5.4: The applied pulsed field amplitudes and the time for extraction of the trapped fields in the simulation

Temperature (K)	Pulsed field amplitude (T)	Time of measurement (s)
77	1.3	43
55	2.5	71
35	2.9	105
15	3.1	175

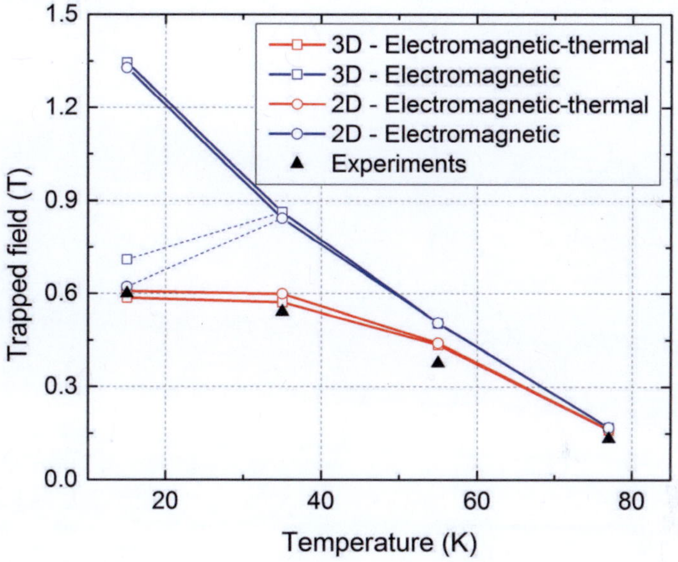

Figure 5.35: The trapped fields of the stack magnetized by PFM with the equilibrium temperatures calculated by numerical simulations and corresponding reported experimental results [84]. The results obtained with four models are compared including 3D and 2D models with or without coupling the heat equation. The models use applied fields and relaxing time in Tab. 5.4. The 2D and 3D electromagnetic models use applied field of 5 T at 15 K because 3.1 T is not enough for fully magnetizing the stack when thermal effect is not considered. The calculated trapped fields obtained by applied field of 3.1 T are kept for reference as dash lines.

As shown in Fig. 5.35, the trapped fields obtained by electromagnetic models are much higher than those obtained by electromagnetic-thermal coupled models. When the thermal effects are not considered, the calculated trapped fields are corresponding to the field cooling or zero field cooling when the temperature is constant through the process.

For modelling of PFM process, the temperature variation has to be considered. In comparison with the measured trapped fields, both 3D and 2D electromagnetic-thermal coupled model match the measurements within 20 %. The numerical models provide a good estimation of the trapped fields, although there are numerous uncertainties related to material properties and cooling conditions, which also vary greatly with temperature.

The trapped fields obtained by 2D and 3D models are very close; however, the current and temperature distribution are quite different during the process. For comparison, as shown in Fig. 5.37 and Fig. 5.38, the current distributions and temperature distributions calculated with 3D and 2D electromagnetic-thermal coupled models during the magnetization at the time points of Fig. 5.36 at 55 K are compared. For simplification, only one fourth of the central flat plane (x>0 and y > 0, z = 0 in Fig. 5.30) of the stack is shown.

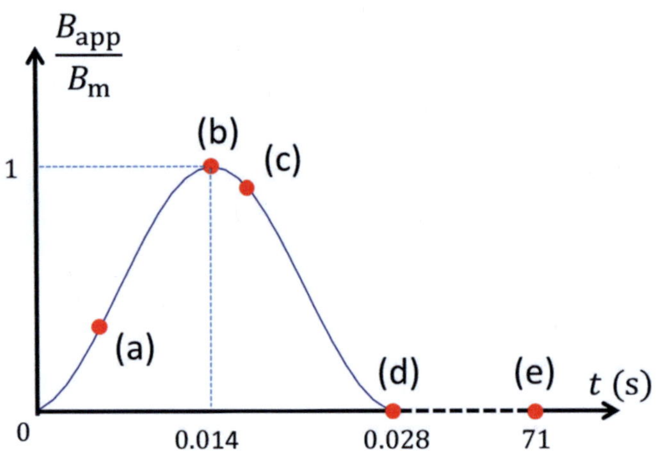

Figure 5.36: The shape of the applied pulsed field as described by Eq. 5.7. The marked time points are used for presenting the current and temperature distribution during the magnetization for Fig. 5.37 and Fig. 5.38. (a) 0.005 s; (b) 0.014 s; (c) 0.016 s; (d) 0.028 s; (e) 71 s.

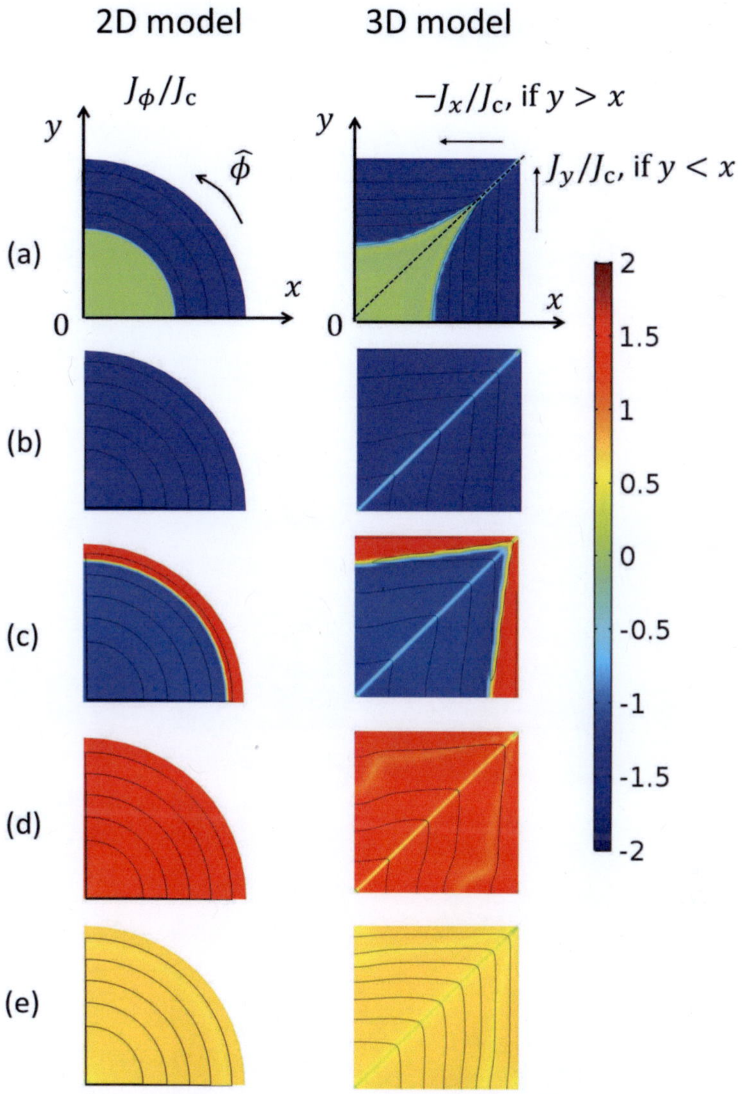

Figure 5.37: The normalized current density distributions and current steam lines for one fourth of the central flat plane (x, y > 0, z = 0 in Fig. 5.30) obtained by 2D and 3D electromagnetic-thermal coupled models during the magnetization for time points as marked in Fig. 5.36. For the 2D model, the color map shows the tangential current density J_ϕ normalized by J_c, since the current distribution is circular. For the 3D model, the color map shows the current density along a square route in the counter-clock direction ($-J_x/J_c$ if y>x and J_y/J_c if x>y). The 2D model is axisymmetric. The plot is obtained by rotation of the axisymmetric plane for better comparison with the 3D model.

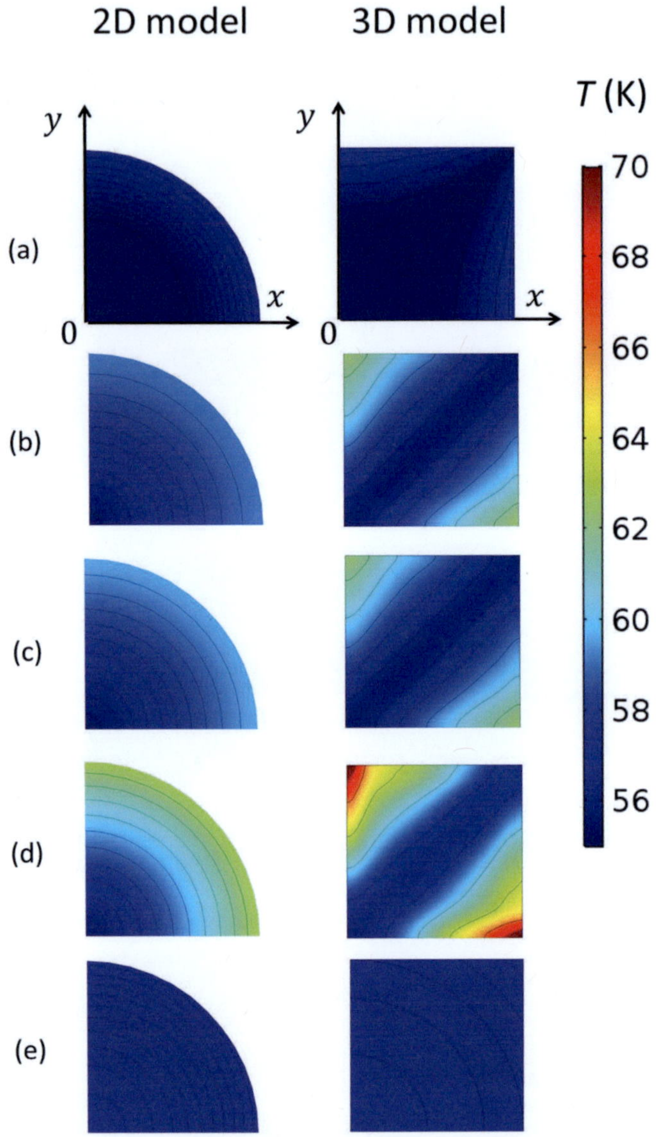

Figure 5.38: The temperature distributions and contour lines for one fourth of the central flat plane (x, y > 0, z = 0 in Fig. 5.30) obtained by 2D and 3D electromagnetic-thermal coupled models during the magnetization for time points as marked in Fig. 5.36. The 2D model is axisymmetric. The plot is obtained by rotation of the axisymmetric plane for better comparison with the 3D model.

As shown in Fig. 5.37, the current distribution obtained by the axisymmetric 2D model is circular in the flat plane. The current penetrates uniformly along the perimeter, corresponding to the situation of a cylinder bulk. The 3D model accounts for the realistic geometry of the square cuboid stack. The current stream lines follow the square boundary of the stack. Finally, the remnant currents distribute in circles and squares for the 2D and 3D models, respectively, as shown in Fig. 5.37(e).

Relating to the penetration process, the temperature distributions obtained by the two models are completely different as shown in Fig. 5.38. For the 2D model, the temperature distributes axisymmetrically and the temperature contour lines are circular, because the heat is generated uniformly along the perimeter. For the 3D model, the current penetrates as square loops following the geometry and more heat is generated on the edges of the square as shown in Fig. 5.38(d). Right after the magnetization at 0.028 s, the maximum temperature is 63 K for the 2D model and 71 K for the 3D model.

The 3D model provides a more realistic current and temperature distribution than the 2D model for a cuboid stack, which cannot be reliably reduced to 2D. However, the 3D model is more time consuming and numerical accuracy is compromised. In this case, the 2D model has a much denser mesh compared to the 3D model, while the 2D model takes less than 1 hour to run and the 3D model takes around 3 hours.

At 55 K, the trapped fields with time calculated with 3D and 2D models are compared in Fig. 5.39. For both the electromagnetic models and the electromagnetic-thermal coupled models, the trapped fields calculated with 3D and 2D models are quite close through the whole process of PFM and come to similar final trapped fields. In conclusion, for simply evaluating the trapped fields produced by PFM, the 2D model can provide a fast and fair estimation.

5.2.4.2 Comparison of Flat and Curved Stacks

For application in the rotor of an electrical machine, a curved stack is geometrically more applicable than a common flat stack. The trapped field distribution produced by field cooling has already been calculated with static model in Chapter 5.2.3. In this chapter, the curved stack magnetized by PFM is simulated with 3D electromagnetic-thermal coupled model and compared with a flat stack. The construction of the model has been introduced in Chapter 5.2.1 and 5.2.2.

As shown in Fig. 5.40, the trapped field as a function of the amplitude of the applied field shows a similar tendency for the flat and curved stacks. The trapped field has a maximum value when the applied field is 2 T; otherwise, the stack is either under-magnetized or over-magnetized. The difference of the trapped fields between the curved and flat stacks is only around 10 %.

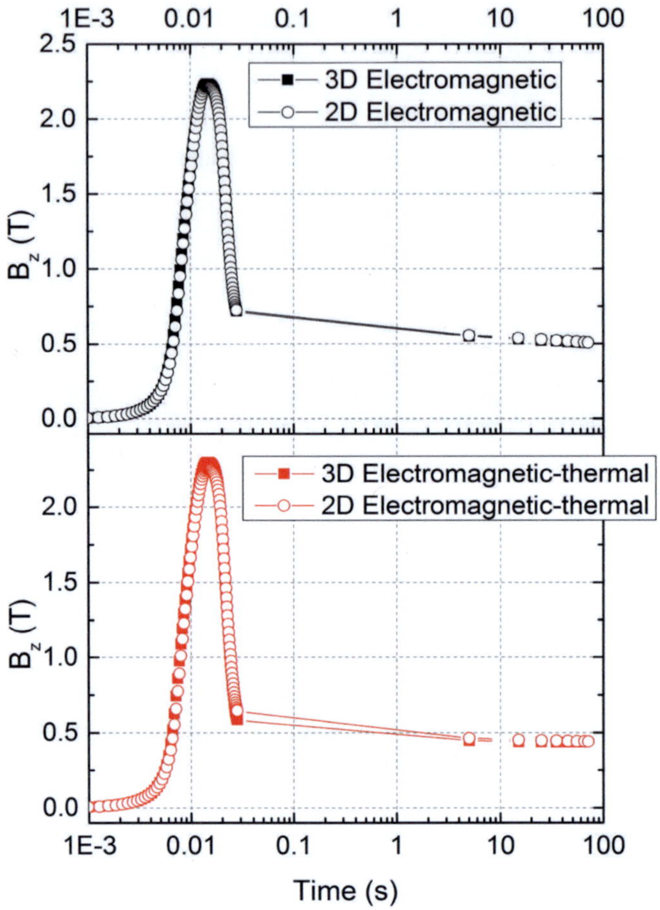

Figure 5.39: The trapped fields with time for 55 K calculated by 3D and 2D models are compared. The upper and lower plots show results for models without and with coupling the heat equation, respectively.

To visualize the difference caused by the bending of the stack, the normalized current distributions on the stack surface for the flat and curved stacks are compared in Fig. 5.41. The amplitude of the applied field is 3 T. The selected time points refer to Fig. 5.36. Only one eighth of flat stack ($x > 0$, $y > 0$ and $z > 0$ in Fig. 5.29(a)) and one fourth of the curved stack ($x>0$ and $y > 0$ in Fig. 5.29(b)) are shown for simplification. As shown in Fig. 5.36, the penetration processes of the flat and curved stacks are similar despite the curvature. The currents flow in square loops. For the curved stack, the currents in the x-z plane only flow in the tangential direction of the curve as constrained by Eq. 5.6. The current distribution of the curved stack is changed with the geometrical deformation and shows no essential difference from that of a flat stack.

The remnant magnetic field distributions 0.8 mm above the stack surface along the x and y axis of the flat and curved stacks at time point (e) in Fig. 5.36 are compared in Fig. 5.42. For the flat stack, the magnetic field distributions along the x and y axis are identical due to the symmetry of the square. For the curved stack, the magnetic field distributions along the two directions are different. The curved stack is curved around the y axis as shown in Fig. 5.29(b), which means that the width of the stack along the x axis is shortened. The field distribution along the y axis is similar to that of the flat stack. The field distribution along the x axis, however, presents a narrower peak compared to that along the y axis of the flat stack. The magnetic field on the periphery ($5 < |x| < 10$ mm) is zero instead of being negative as in the flat stack.

Generally, based on the performed simulations, the curved stack presents no essential difference compared to the flat stack. The remnant magnetic field distribution along the curved direction is changed due to the curvature, which will be further justified by the measurements in Chapter 5.3.1 and 5.3.2.2.

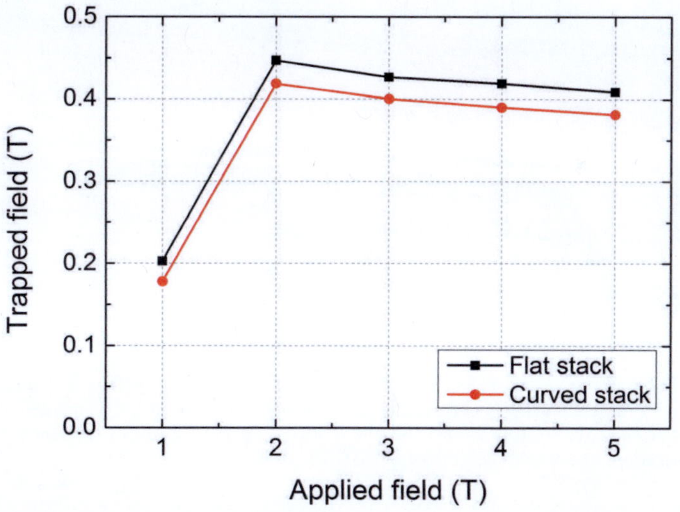

Figure 5.40: The trapped field with the amplitude of the applied field for the flat stack and the curved stack calculated with 3D simulations. The trapped field is extracted 0.8 mm above the surface center of the of the stack 71 s after the pulse when the temperature has recovered to 55 K.

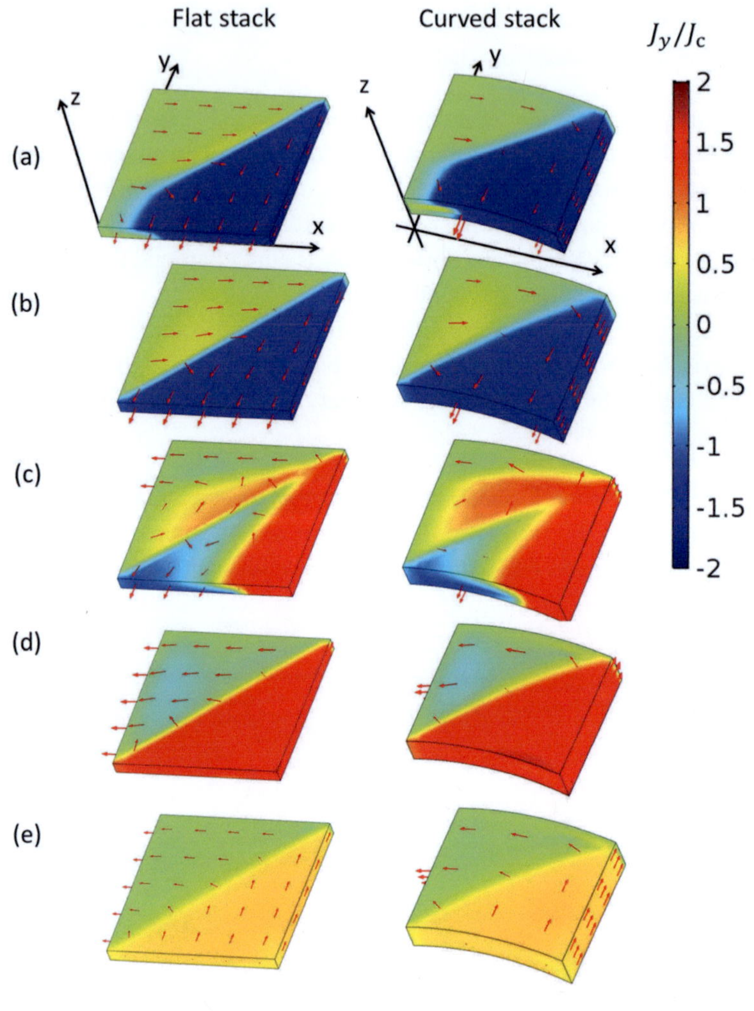

Figure 5.41: The normalized current density and current arrows on the stack surface. The color map shows the current density along the x direction J_x normalized by J_c. The selected points are as marked in Fig. 5.36. Only one eighth of flat stack (x, y, z > 0 in 2.1(a)) and one fourth of the curved stack (x>0 and y > 0 in 2.1(b)) are shown for simplification. The amplitude of the applied field is 3 T.

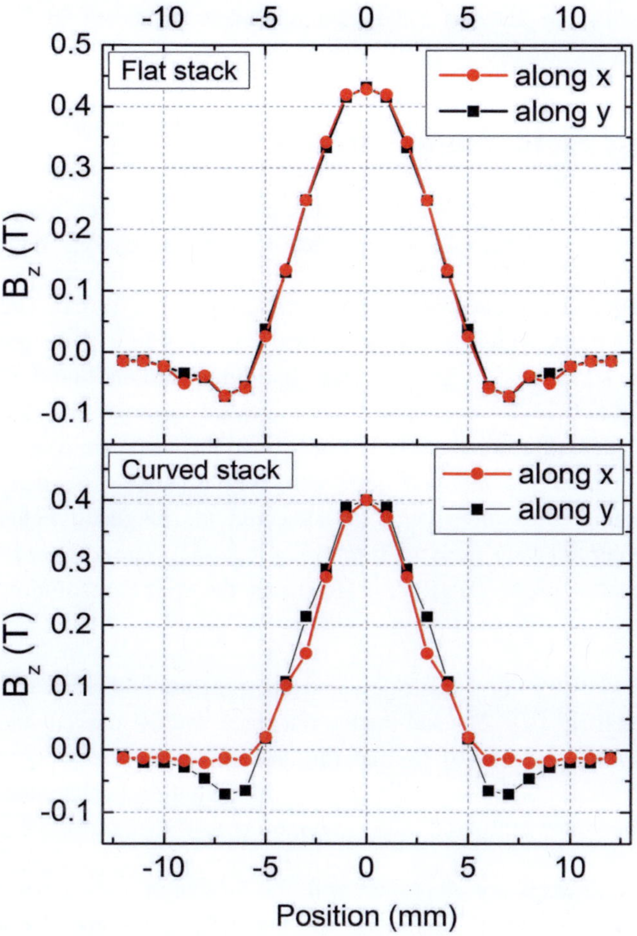

Figure 5.42: The trapped field B_z distributions along the x and y directions of the flat and curved stacks extracted 0.8 mm above the stacks. The amplitude of the applied field is 3 T and the field distributions are extracted 71 s after the pulse, corresponding to the case (e) in Fig. 5.41.

5.3 Experiments

Magnetization experiments were carried out on HTS stacks. The static field magnetization experiments in Chapter 5.3.1 are compared with the 3D static model described in Chapter 5.2.3. The pulsed field magnetization experiments in Chapter 5.3.2

are compared with the 2D and 3D dynamic models presented in Chapter 5.1.4 and Chapter 5.2.4.2, respectively.

5.3.1 Static Field Magnetization

The magnetic field distributions of a flat stack and a curved stack magnetized by static field magnetization were mapped and compared with the simulations in Chapter 5.2.3.

The tape is produced by Superpower and the brand type of SF12030. The tape is 12 mm wide and 35 μm thick with a self-field critical current 450 A. The tape was cut into 12 mm ×12 mm squares. The sample was first immersed in liquid nitrogen, then a magnetic field up to 0.11 T was supplied and decreased to zero by an electromagnet. The magnetic field distribution was scanned 1 mm above the surface of the stack (z = 1 mm in Fig. 5.29(a)). Then the identical tape was bent around a wooden rod of 20 mm diameter. The same procedures were repeated and the magnetic fields were mapped 1 mm above the curved stack (z = 11 mm in Fig. 5.29(b)). The magnetic field is mapped with an Arepoc Hall probe (HHP-VA) [166] and the spatial resolution of the scan is 0.3 mm.

The magnetic field distribution was calculated using the model described in Chapter 5.2.3. The criterion electrical field E_c^* used for the flat stack and the curved stack are 4 nV and 0.3 nV, respectively, obtained by normalizing the calculated peak magnetic fields equal to the measured values when n equals 21. The measured and calculated magnetic field maps for both stacks are compared in Fig. 5.43.

The calculated field maps are consistent with the measured ones for both the flat stack and the curved stack. For the flat stack, the magnetic fields are identical in the four quadrants due to symmetry. The magnetic fields are negative on the periphery regions of the stack. For the curved stack, the magnetic field distribution along the y axis is similar to that of flat stacks; however, the peak width along the x axis is smaller and negative fields are not present.

The magnetic field distributions along x and y axis (along dash lines in Fig. 5.43) are extracted for better comparison as shown in Fig. 5.44. For the flat stack, the calculated field distributions along x and y axis are the same due to symmetry; however, the measured field peak along the y axis is a bit narrower than that along the x axis. The reason is possibly that the length of the cut sample along the y axis is a bit shorter than the planned 12 mm. Another possible reason is that the tape is anisotropic along the directions of length and width due to doping [34]. For the curved stack, the width of the field peak along the x axis is smaller than that along the y axis as shown by the simulation. The fields on the periphery along the x axis ($5 < |x| < 10$ mm) are close to zero instead of being negative due to the curvature. For practical applications, the curved stack provides a lower peak field compared to the flat one, suggesting that more materials should be used in order to keep the same magnetic flux density using curved stacks. Generally, the static 3D modelling provides a proper explanation for the experimental results with little computation time. This static 3D model only takes several seconds to run, but a 3D dynamic model with the same discretization takes several hours.

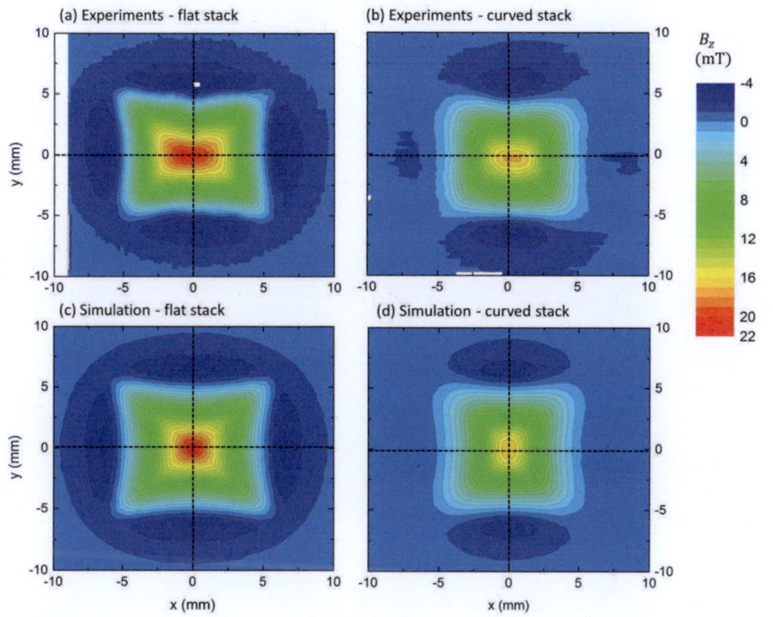

Figure 5.43: The magnetic flux density B_z maps measured 1 mm above the surface of a single layer of (a) flat stack and (b) curved stack; and corresponding $\boldsymbol{B_z}$ maps calculated by simulations for (c) the flat stack and (d) the curved stack. The curved stack is bended around the y axis as shown in Fig. 5.29(b).

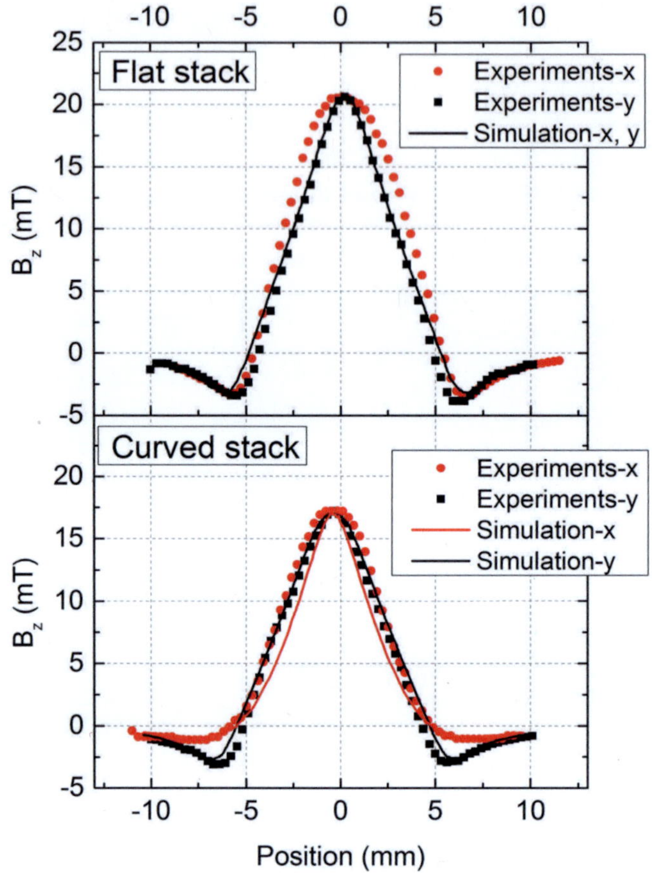

Figure 5.44: The magnetic field distributions along x and y directions extracted along the dash lines in Fig. 5.43. The experimental and numerical results are compared for the flat stack and the curved stack, respectively.

5.3.2 Pulsed Field Magnetization

In this chapter, experiments on flat stacks with multi-pulse technique in Chapter 5.3.2.1 is used to validate the proposed strategy by the 2D modelling presented in Chapter 5.1.4; and experiments on both flat and curved stacks in Chapter 5.3.2.2 are compared to the 3D dynamic modelling presented in Chapter 5.2.4.2.

5.3.2.1 Flat Stack Magnetized with Multi-Pulse Technique

In order to validate the simulation for magnetization of HTS stacks with multi-pulse technique in presented Chapter 5.1.4, experiments were carried out on a stack of 20-layer coated conductors. The tape was 12 mm wide produced by SuperPower Inc. It was cut into 12 mm ×12 mm squares and tightly piled together. More information about the tapes can be found in [144]. The stack was magnetized by PFM at 30 K. The experimental setups are the same as those reported in [58], [73], [84].

The trapped field was measured 0.8 mm above the surface center of the stack 30 s after each applied pulse. After each pulse sequence, the stack was heated up to 95 K to ensure that the sample was clean of currents, so every tested sequence was started when there were no currents in the sample. Every pulse sequence was tested for three times with the same conditions on different dates to test for reproducibility and possible sample degradation. Corresponding to Fig. 5.25 and 5.26, the pulse sequences of different starting amplitudes or different amplitude intervals were tested, as shown in Fig. 5.45 and 5.46, respectively.

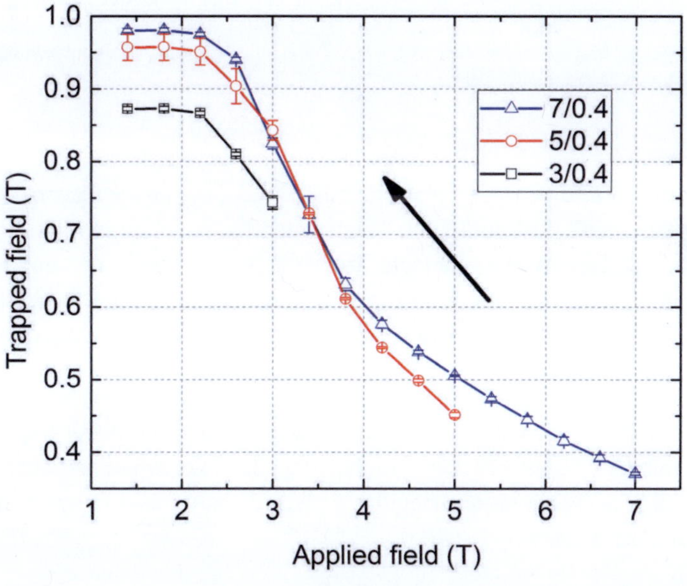

Figure 5.45: The experimental results of the trapped fields of a stack magnetized by pulse sequences with different starting amplitudes. "7/0.4" denotes a pulse sequence which starts from 7 T and reduces the amplitude by 0.4 T for each pulse. And so on, for other sequences. The trapped field and flux evolve along each line from the right to the left with each applied pulse.

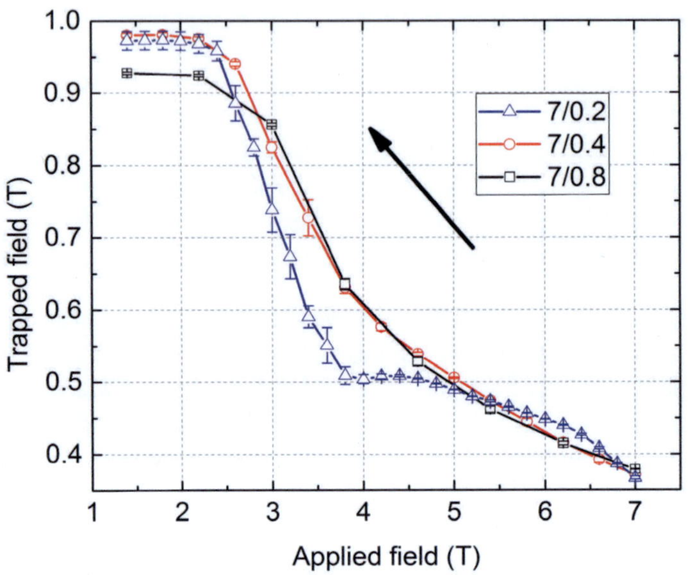

Figure 5.46: The experimental results of the trapped fields of a stack magnetized by pulse sequences with different amplitude intervals. "7/0.4" denotes a pulse sequence which starts from 7 T and reduces the amplitude by 0.4 T for each pulse. And so on, for other sequences. The trapped field and flux evolve along each line from the right to the left with each applied pulse.

As shown in Fig. 5.45, the larger the starting amplitude, the larger the final trapped field, which is consistent with the simulation. The trapped fields are quite reproducible at large applied fields, but they start to deviate at about 3-4 T where the trapped fields are increasing rapidly. This may relate to the flux jumps which cannot be predicted by the numerical model in this work. A more accurate model has to be developed to explain this.

The pulse sequences that start from 7 T and reduce the amplitude with different intervals are compared in Fig. 5.46. Comparing "7/0.4" and "7/0.8", the pulse sequence with the smaller field interval "7/0.4" reaches a higher trapped field, which is consistent with the simulation. "7/0.2" shows a rather irregular pattern. It may relate to some instability that we cannot understand yet.

The deviations between the simulation and experiments include several aspects. The laminated structure and the material properties in the simulation are not the same for the sample in the experiments. The simulation considers an infinitely long stack because long stacks are more favorable for practical applications. In the experiments, the sample is a small cuboid which fits for the experimental set-ups. The thermal conditions in the experiments are difficult to judge. Despite these issues, the experiments lead to the same optimal applied sequences for PFM as described in Chapter 5.1.4.2, which justifies the strategy proposed by the simulation.

5.3.2.2 Curved Stack Magnetized by Pulsed Field Magnetization

In this chapter, the influence of curvature on stacks magnetized by PFM is investigated experimentally for validation of the simulations presented in Chapter 5.2.4.2. The experimental setups are the same as those reported in [58], [73], [84]. The sample was made from a 12-mm-wide tape produced by SuperPower Inc., of the brand type SF12030. The nominal critical current was 450 A. The tape was based on 30 μm substrate with 3 μm silver stabilizer. It was cut into 12 mm×12 mm squares and tightly piled together to make a 20-layer stack.

The stack was magnetized by PFM with the IMRA technique (iteratively magnetizing pulsed field method with reducing amplitude). Two sets of experiments were carried out separately at 30 K and 60 K. After magnetization and measurement of each pulse sequence, the sample was heated up to 95 K to remove current from previous experiments. For 30 K, the pulse sequence started from a pulse of magnitude 7 T and reduced by 0.4 T for each applied pulse. For 60 K, the pulse sequence started from a pulse of the magnitude 4.2 T. The magnetic field distribution was measured 30 s after each applied pulse 0.8 mm above the surface of the stack along the x axis as shown in Fig. 5.29(b).

The measured magnetic field distributions are shown in Fig. 5.47 and 5.48 for 30 K and 60 K, respectively. To make a fair comparison between flat and curved stacks, the sample was first magnetized and measured as a flat stack, which is shown in the upper plot as "flat stack before". Then, the stack was curved around the y axis by tightly wrapping it around a wooden rod of 20 mm diameter as shown in Fig. 5.29(b). The curved stack was magnetized and measured again, the results of which are plotted in the middle of Fig. 5.47 and 5.48 as "curved stack". In the end, the stack was recovered to be flat and the measurement was repeated as shown in the bottom plot as "flat stack after".

For both 30 K and 60 K, the magnetic field distributions of "flat stack before" show a regular pattern. The magnetic field distribution presents a conical shape and grows higher with after each applied pulse, consistent with experiments and modelling for both HTS bulks and stacks [2], [58], [73], [84], [85].

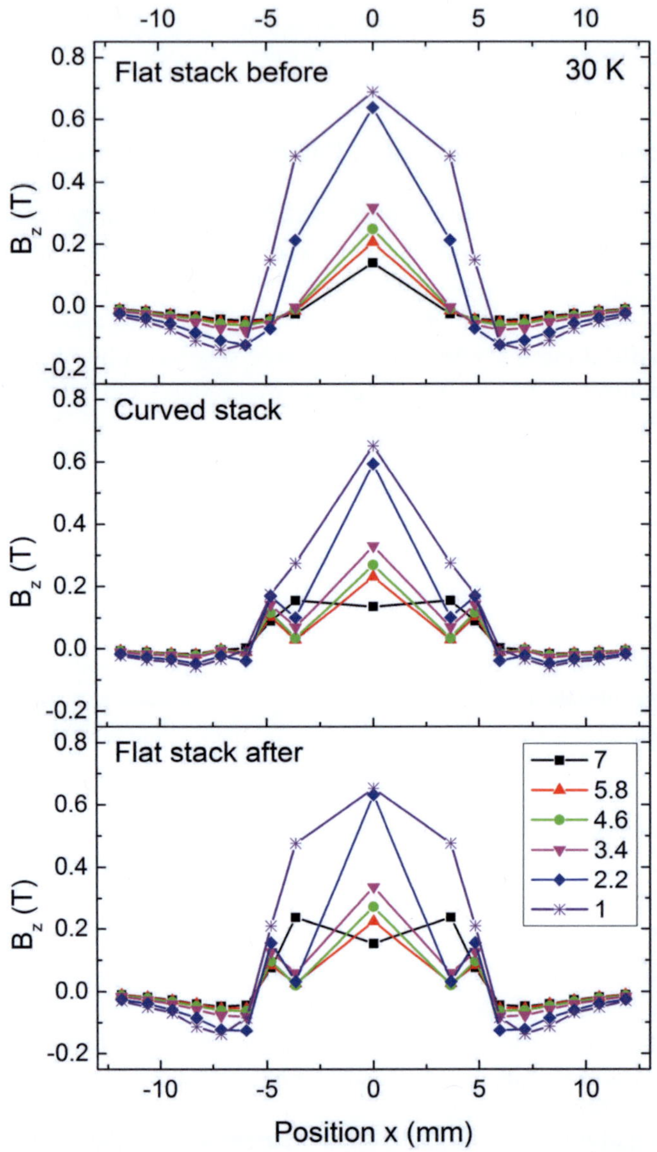

Figure 5.47: The measured field distribution 0.8 mm above the surface of the curved stack along the x direction in Fig. 5.29(b) at 30 K. The stack was magnetized by a pulse sequence starting with amplitude 7 T and reducing the amplitude by 0.4 T for each pulse. Every third pulse is presented for clarity in the visualization. "Flat stack before": the stack was first magnetized and measured as a flat stack; "Curved stack": the same stack was then curved and the measurement was repeated; "Flat stack after": the stack was recovered flat and the measurement was repeated.

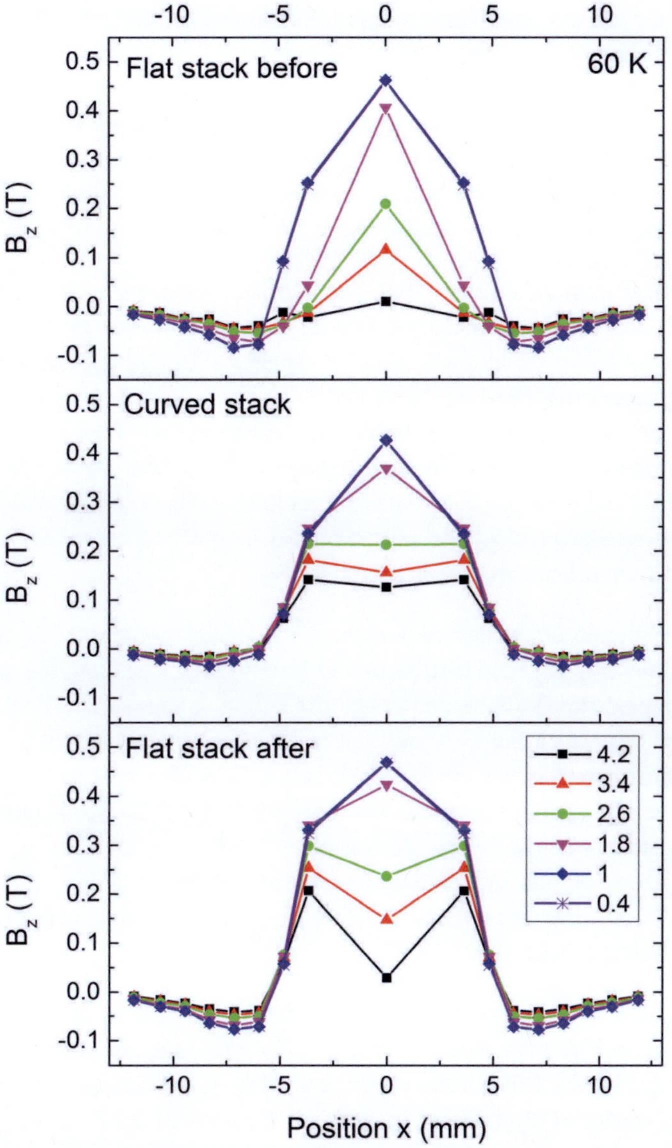

Figure 5.48: The measured field distribution 0.8 mm above the surface of the curved stack along the x direction in Fig. 5.29(b) at 60 K. The stack was magnetized by a pulse sequence starting with amplitude 4.2 T and reducing the amplitude by 0.4 T for each pulse. Every other pulse is presented for clarity in the visualization. "Flat stack before": the stack was first magnetized and measured as a flat stack; "Curved stack": the same stack was then curved and the measurement was repeated; "Flat stack after": the stack was recovered flat and the measurement was repeated.

The results of the curved stack, however, present an unusual pattern. After application of several initial pulses of large amplitudes, the magnetic field distribution presents an M-shaped profile. The central field grows with application of successive fields and finally builds into a conical field distribution. The evolution pattern of the field distribution cannot be explained with the model in Chapter 5.2.4.2. According to the simulation, the evolution pattern of the field distribution of the curved stack should be similar to that of the flat stack, which presents a conical shape at all times.

The further measurement of the "flat stack after", presents a transition from the M-shape to the conical-shape distributions with successive pulses as well, similar to the "curved stack" situation. "Flat stack after" presents a different pattern from "flat stack before", which means that PFM on the curved stack has produced some irreversible effect on the stack. This influence produced by the bending operation is not a degradation of the critical current density, because the final trapped field of "flat stack after" is equal or even a bit higher than that of the "flat stack before". The experiments were repeated twice and gave similar results. The influence and its mechanism cannot be understood based on the current scheme of numerical modelling.

Although the evolution tendency was somehow influenced by the PFM on the bending of the stack, the final trapped field distribution of both the flat stack and the curved stack is of a normal conical shape as the numerical simulation suggests. The final trapped field distributions of "flat stack before" and "flat stack after" are almost the same. The final trapped field distribution of the curved stack is conical with a smaller width of the peak than that of the flat stack. The fields on the periphery of the stack along the x axis ($5 < |x| < 10$ mm) are close to zero due to the curvature, instead of being negative as in a flat stack. These results are consistent with the dynamic simulation presented in Chapter 5.2.4.2 and also the static simulation and field mapping measurement results described in Chapter 5.3.1.

Further study has to been carried out with different tapes to fully understand the irreversible influence on the curved stack caused by PFM. Nevertheless, the final trapped field of the curved stack follows the prediction of the numerical simulation, suggesting that the curved stack can be produced by PFM with empirical pulse sequences to produce a conical trapped field with no obvious degradation compared to a flat one. In this sense, curved stacks are applicable for replacing ordinary flat stacks in practical applications with no need to modify the original PFM system.

5.4 Summary

This chapter has investigated the magnetization of stacks of HTS coated conductors numerically and experimentally. For the numerical simulation, 2D and 3D models have been developed.

For the 2D modelling in Chapter 5.1, a 2D planar electromagnetic-thermal coupled model is developed for describing an infinitely long stack of HTS coated conductors magnetized by pulsed field magnetization (PFM). The model is characterized by considering realistic laminated structures of HTS coated conductors. Temperature and anisotropic magnetic field dependences of the critical current density of HTS and temperature dependence of other properties of all composing materials including electrical resistivity, thermal conductivity and heat capacity are considered. Based on the model, two strategies to improve the trapped field produced by PFM are suggested. Chapter 5.1.3 suggests using an arrangement of controlled magnetic density distribution coils instead of a common solenoid. The roles of different composing materials are discussed. Chapter 5.1.4 suggests that optimal applied pulse sequences for PFM are those of descending amplitudes with small amplitude intervals.

For the 3D modelling in Chapter 5.2, 3D models are developed to describe the magnetization of stacks of HTS coated conductors, including an ordinary flat cuboid stack and a curved stack that fit for application in a rotor. In Chapter 5.2.3, a static 3D model is developed to describe the flat and curved stacks magnetized by static field magnetization with a prescribed current distribution. In Chapter 5.2.4, an electromagnetic-thermal coupled model extended from Chapter 5.1 is used to describe the stacks magnetized by PFM. The techniques for reducing the computation time are introduced, including homogenization of the stack with constraint of currents and adding jacket layers for reducing mesh elements. Using the model, 2D and 3D simulations of HTS stacks magnetized by PFM are compared with experiments found in the literature, suggesting that 2D models can provide a good estimation of the trapped fields, but 3D models present more accurate current and temperature distributions. 3D simulations of flat and curved stacks are compared and show that the trapped field distribution along the curved edge of the curved stack is changed due to the bending.

Finally in Chapter 5.3, experiments are carried out to validate the previous simulations. For the static field magnetization in Chapter 5.3.1, the remnant field distributions of a flat and a curved stack magnetized by zero field cooling are mapped and compared with the simulation presented in Chapter 5.2.3. The magnetic field distributions obtained numerically and experimentally agree quantitatively. For the PFM experiments in Chapter 5.3.2.1, a flat square stack is magnetized with different pulse sequences according to the simulations presented in Chapter 5.1.4. The results justify the suggested approach for optimal sequences. In Chapter 5.3.2.2, a flat stack sample is magnetized by PFM and compared to the identical stack after bending and recovering to be flat. The final trapped field distributions agree with the simulations in Chapter 5.2.4.2, showing that the remnant field distributions along the curved direction of the curved stack are changed compared to the flat one. The field distribution patterns of the stack after bending present irregular M-shape field distributions after initial pulses, which are not explainable using the current models, and need further investigation.

6 Conclusion and Perspective

High temperature superconducting (HTS) bulks and stacks of coated conductors can be magnetized to become trapped-field magnets (TFMs), which can provide much stronger static magnetic fields than those provided by conventional permanent magnets. Pulsed field magnetization (PFM) can provide cost-effective, flexible and compact *in situ* magnetization, which is attractive for realizing practical application of trapped-field magnets. The key issue with PFM is that the trapped field produced is generally lower than that produced by the field cooling method, due to the fast flux motion that leads to heat generation and temperature increase. The motivation of the thesis is to investigate the flux dynamics and propose strategies to improve the trapped field produced by PFM by means of numerical simulations with electromagnetic-thermal coupled models and experimental validation.

For HTS bulks, the geometry and structure are simpler and the research in the literature is more extensive compared to stacks of coated conductors. In Chapter 4, the research on HTS bulks is carried out with 2D axisymmetric models and extended from previous publication as basis for the research on stacks of coated conductors. In Chapter 4.2.1, it is found that using the controlled magnetic density distribution coils can further improve the trapped fields produced by PFM compared to the previously reported vortex coils. Consistently with literature, the reason is that the different ways of current penetration by using different coils result in less heat generation during the magnetization. The increase in the trapped field, however, is not as significant as reported by previous literature, which leads to a further discussion about the influence of parameters used in simulations of bulks in Chapter 4.2.2. The exponent parameter n in the E-J power law and the B_0 parameter in the Kim model are found to greatly influence the simulation results. These parameters are temperature dependent and are not easy to be reliably extracted from the characterization of bulks; besides, HTS bulks are generally inhomogeneous due to the growth sector boundaries caused by the production technique. Based on the comparison with experimental evidence in the literature, the current 2D modelling of HTS bulks is problematic. For reasonable quantitative modelling, more accurate characterization of bulks is needed to account for the position-, temperature- and magnetic field-dependent properties. In addition, other new models describing the magnetization process on a smaller scale to account for microscopic effects may also be worth developing.

The main objects of investigation of the thesis are stacks of HTS coated conductors, which have been recently attracting a lot of interest because they are more commercially available, cost effective, flexible in geometry and mechanically robust compared to HTS bulks. The coated conductors are uniform and can be easily and precisely characterized, which allows reliable modelling. The simulations are carried out with 2D and 3D models and validated by experiments. The contributions are listed following the scheme of modelling with the relevant experiments included.

2D simulations of stacks of coated conductors magnetized by PFM are carried out with 2D electromagnetic-thermal coupled models in Chapter 5.1. The model is characterized by considering the realistic laminated structures of coated conductors and temperature-dependent properties of different composing materials. The temperature and anisotropic magnetic field dependences of the critical current density of coated conductors are considered as well. With the developed model, two strategies are proposed to improve the trapped field produced by PFM. First, using controlled magnetic density distribution coils instead of a common solenoid can improve the trapped field and trapped flux as presented in Chapter 5.1.3. The improvement of trapped field is mainly due to the reduction of heat generation in the metallic parts. Second, the optimal applied pulse sequences are suggested for the PFM with multi-pulse technique in Chapter 5.1.4. The optimal applied pulse sequence initiates with a pulse of large enough amplitude and continues with pulses of descending amplitudes, theoretically with infinitely small intervals. The second strategy is qualitatively validated by experiments in Chapter 5.3.2.1.

3D simulations are used to provide more realistic descriptions of stacks of coated conductors that cannot be easily reduced into 2D modelling. Besides a common flat stack that can be modelled as a square cuboid, a curved stack is also considered for its geometrical applicability for a rotor in an electrical machine in Chapter 5.2. Simulations with 3D static models are used to describe the flat and curved stacks magnetized by field cooling in Chapter 5.2.3. The results show that the flat and curved stacks produce similar trapped field distributions with differences in the periphery along the curved edges. The results show agreements with experiments of field cooling in Chapter 5.3.1.

3D simulations of flat and curved stacks of coated conductors magnetized by PFM are carried out with electromagnetic-thermal coupled models. The thesis proposes to homogenize the stacks by constraining the currents to flow parallel to the tape surface. In this way, the number of mesh elements is greatly reduced and computation time is consequently saved. Based on the model, a cuboid flat stack is first simulated for different temperatures with 2D axisymmetric models and 3D models in Chapter 5.2.4.1. The temperatures and current distributions during the magnetization are compared. The 3D model provides a more realistic description; however, 2D and 3D models achieve similar trapped fields, which suggest 2D axisymmetric models can be used to simulate cuboid stacks when one is only interested in trapped fields. The results are also compared with published experimental data and show agreement within 20 % difference.

Then the flat and curved stacks magnetized by PFM are simulated and compared in Chapter 5.2.4.2. The results show that the curved stack presents no essential differences with respect to the flat one, but the current distribution is changed following the geometrical deformation. In order to validate the simulation, a flat stack is magnetized by PFM with multi-pulse technique in Chapter 5.3.2.2. The identical stack is then curved and recovered to repeat the same magnetization for comparison. The final trapped fields of the flat and curved stacks in experiments show agreement with the 3D simulations. However, the bending of the stack produces some irreversible influence on the stack which results in M-shaped field distributions after initial pulses. Further research is needed to understand this behavior.

Many questions remain for future research on the trapped-field magnets magnetized by PFM. Based on the results of this thesis, three main directions for modelling stacks of coated conductors can be identified:

1. In order to make more precise descriptions and predictions of the experiments, the parameters and assumptions of the models can be further investigated. For example, the temperature and magnetic field dependences of the n value in the E-J power law have not been considered in this work. It is worth including such dependences or assessing the influence of the assumption of constant n values. In addition, the cooling conditions for simulations can be better judged if the temperature measurement is available.

2. The 3D modelling is worth further development. The model used in the thesis considers a stack with no metallic stabilizers and uses a homogenization technique. The influence of laminated structures of metallic components can be investigated using realistic 3D models. For the 3D modelling, increasing the accuracy and reducing the computation time are still sources of concern.

3. As far as practical applications of HTS stacks are concerned, further work should still focus on investigating possible strategies to improve the trapped field produced by PFM. The multi-pulse technique with modified temperatures is worth modelling as the next step to make meaningful suggestions for experimental practice. Other strategies, including different coil configurations and composite bulks with ferromagnetic materials, also remain to be investigated.

Bibliography

[1] J. H. Durrell, A. R. Dennis, J. Jaroszynski *et al.*, "A trapped field of 17.6 T in melt-processed, bulk Gd-Ba-Cu-O reinforced with shrink-fit steel," *Supercond. Sci. Technol.*, vol. 27, no. 8, pp. 082001, Jun. 2014.

[2] M. D. Ainslie and H. Fujishiro, "Modelling of bulk superconductor magnetization", *Supercond. Sci. Technol.*, vol. 28, no. 5, pp. 053002, Mar. 2015.

[3] H. K. Onnes, "Further experiments with Liquid Helium G. On the electrical resistance of Pure Metals etc. VI. On the Sudden Change in the Rate at which the Resistance of Mercury Disappears," *KNAW Proceedings 14 II. Amsterdam Huygens Institute - Royal Netherlands Academy of Arts and Sciences*, pp. 818-821, 1911-1912.

[4] A. C. Rose-Innes and E. H. Rhoderick, "Introduction to Superconductivity," Pergamon Press, 1978.

[5] M. N. Wilson. "Superconducting Magnets," Clarendon Press Oxford, 1983.

[6] M. Meissner and R. Ochsenfeld, "Ein neuer Effekt bei Eintritt der Supraleitfhigkeit," *Naturwissenschaften*, vol. 21, no. 44, pp. 787-788, Nov. 1933.

[7] V. L. Ginzburg and L. D. Landau, "To the theory of superconductivity," *Zh. Eksp. Teor. Fiz.,* vol. 20, pp. 1064-1082, 1950.

[8] A. A. Abrikosov, "On the magnetic properties of superconductors of the second group," *Soviet Phys. JETP-USSR*, vol. 5, no. 6, pp. 1174-1183, 1957.

[9] J. G. Bednorz, and K. A. Müller, "Possible high-Tc superconductivity in the Ba-La-Cu-O system." *Zeitschrift für Physik B Condensed Matter*, vol. 64, iss. 2, pp. 189-193, Apr. 1986.

[10] D. Larbalestier, A. Gurevich, D. Feldmann *et al.*, "High-Tc superconducting materials for electric power applications," *Nature*, vol. 414, pp. 368-377, Nov. 2001.

[11] J. Nagamatsu, N. Nakagawa, T. Muranaka *et al.*, "Superconductivity at 39 K in magnesium diboride," *Nature*, vol. 410, pp. 63-64, Mar. 2001.

[12] Y. Kamihara, H. Hiramatsu, M. Hirano *et al.*, "Iron-based layered superconductor: LaOFeP," *J. Am. Chem. Soc.*, vol. 128, no. 31, pp. 10012-10013, Jul. 2006.

[13] Y. Kamihara, T. Watanabe, M. Hirano *et al.*, "Iron-based layered superconductor La[$O_{1-x}F_x$]FeAs (x = 0.05–0.12) with T_c = 26 K," *J. Am. Chem. Soc.*, vol. 130, no. 11, pp. 3296-3297, Feb. 2008.

[14] I. Pallecchi, M. Eisterer, A. Malagoli *et al.*, "Application potential of Fe-based superconductors", *Supercond. Sci. Technol.*, vol. 28, no. 11, pp. 114005, Oct. 2015.

[15] A. Ishihara, T. Akasaka, M. Tomita *et al.*, "Superior homogeneity of trapped magnetic field in superconducting MgB_2 bulk magnets," *Supercond. Sci. Technol.*, vol. 30, no. 3, pp. 035006, Jan. 2017.

[16] J. Zou, M. D. Ainslie, H. Fujishiro *et al.*, "Numerical modelling and comparison of MgB_2 bulks fabricated by HIP and infiltration growth," *Supercond. Sci. Technol.*, vol. 28, no. 7, pp. 075009, May 2015.

[17] H. Fujishiro, H. Mochizuki, M. D. Ainslie *et al.*, "Trapped field of 1.1 T without flux jumps in an MgB_2 bulk during pulsed field magnetization using a split coil with a soft iron yoke," *Supercond. Sci. Technol.*, vol. 29, no. 8, pp. 084001, Jun. 2016.

[18] J. D. Weiss, A. Yamamoto, A. A. Polyanskii *et al.*, "Demonstration of an iron-pnictide bulk superconducting magnet capable of trapping over 1 T," *Supercond. Sci. Technol.*, vol. 28, no. 11, pp. 112001, Sep. 2015.

[19] K. Sawano, M. Morita, M. Tanaka *et al.*, "High magnetic flux trapping by melt-grown YBaCuO superconductors," *Jpn. J. Appl. Phys.*, vol. 30, no. 7A, part 2, pp. L1157-L1159, Jul. 1991.

[20] Y. Shi, J. H. Durrell, A. R. Dennis *et al.*, "Multiple seeding for the growth of bulk GdBCO–Ag superconductors with single grain behaviour," *Supercond. Sci. Technol.*, vol. 30, no. 1, pp. 015003, Nov. 2016.

[21] M. D. Ainslie, H. Fujishiro, T. Ujiie *et al.*, "Modelling and comparison of trapped fields in ReBCO bulk superconductors for activation using pulsed field magnetization," *Supercond. Sci. Technol.*, vol. 27, no. 6, pp. 065008, Apr. 2014.

[22] M. Tomita and M. Murakami, "Improvement of the mechanical properties of bulk superconductors with resin impregnation," *Supercond. Sci. Technol.*, vol. 13, no. 6, pp. 722-724, Feb. 2000.

[23] S. B. Kim, J. Matsunaga, Y. Fujii *et al.*, "The levitation characteristics of ferromagnetic materials by ring-shaped HTS bulks with two trapped field distributions," *IEEE Trans. Appl. Supercond.*, vol. 23, no. 3, pp. 4603204, Jun. 2013.

[24] S. Nariki, N. Sakai, M. Murakami *et al.*, "High critical current density in Y–Ba–Cu–O bulk superconductors with very fine Y211 particles," *Supercond. Sci. Technol.*, vol. 17, no. 2, pp. S30-S35, Jan. 2004.

[25] M. Carrera, X. Granados, J. Amoros *et al.*, "Detection of current distribution in bulk samples with artificial defects from inversion of Hall magnetic maps," *IEEE Trans. Appl. Supercond.*, vol. 19, no. 3, pp. 3553-3556, Jun. 2009.

[26] http://www.superpower-inc.com/content/2g-hts-wire

[27] A. Sundaram, Y. Zhang, A. R. Knoll *et al.*, "2G HTS wires made on 30 μm thick Hastelloy substrate," *Supercond. Sci. Technol.*, vol. 29, no. 10, pp. 104007, Sep. 2016.

[28] Y. B. Kim, C. F. Hempstead and A. R. Strnad, "Critical persistent currents in hard superconductors," *Phys. Rev. Lett.*, vol. 9, no. 7, pp. 306-309, Oct. 1962.

[29] P. W. Anderson, "Theory of flux creep in hard superconductors," *Phys. Rev. Lett.*, vol. 9, no. 7, pp. 309-311, Oct. 1962.

[30] M. Jirsa, L. Pust, and D. Dlouhy, "Fishtail shape in the magnetic hysteresis loop for superconductors: Interplay between different pinning mechanisms," *Phys. Rev. B*, vol. 55, no. 5, pp. 3276-3284, Feb. 1997.

[31] T. H. Johansen, M. R. Koblischka, H. Bratsberg *et al.*, "Critical-state model with a secondary high-field peak in $J_c(B)$," *Phys. Rev. B*, vol. 56, no. 17, pp. 11273-11278 Nov. 1997.

[32] F. Sirois and F. Grilli, "Potential and limits of numerical modelling for supporting the development of HTS devices", *Supercond. Sci. Technol.*, vol. 28, no. 4, pp. 043002, Mar. 2015.

[33] F. Grilli, E. Pardo, A. Stenvall *et al.*, "Computation of losses in HTS under the action of varying magnetic fields and currents," *IEEE Trans. Appl. Supercond.*, vol. 24, no. 1, pp. 8200433, Feb. 2014.

[34] F. Grilli, F. Sirois, V. M. R. Zermeño *et al* "Self-consistent modeling of the Ic of HTS devices: how accurate do models really need to be?" *IEEE Trans. Appl. Supercond.*, vol. 24, no. 6, pp. 8000508, Dec. 2014.

[35] http://www.victoria.ac.nz/robinson/hts-wire-database

[36] A. M. Campbell and J. E. Evetts, "Flux vortices and transport currents in type II superconductors," *Adv. Phys.*, vol. 21, iss. 90, pp. 199-428, 1972.

[37] Y. Yeshurun and A. P. Malozemoff, "Giant flux creep and irreversibility in an Y-Ba-Cu-0 crystal: An alternative to the superconducting- glass model," *Phys. Rev. Lett.*, vol. 60, no. 21, pp. 2202-2205, May 1988.

[38] C. P. Bean, "Magnetization of hard superconductors," *Phy. Rev. Lett.*, vol. 8, no. 6, pp. 250-253, Mar. 1962.

[39] C. P. Bean, "Magnetization of high-field superconductors," *Rev. Mod. Phys.*, vol. 36, no. 1, pp. 31-39, Jan. 1964.

[40] C. Navau, N. Del-Valle, and A. Sanchez, "Macroscopic modeling of magnetization and levitation of hard Type-II superconductors: the critical-state model," *IEEE Trans. Appl. Supercond.*, vol. 23, no. 1, pp. 8201023, Feb. 2013.

[41] J. Rhyner, "Magnetic properties and AC-losses of superconductors with power law current-voltage characteristics," *Physica C*, vol. 212, iss. 3-4, pp. 292-230, Jul. 1993.

[42] https://webstore.iec.ch/preview/info_iec61788-3%7Bed2.0%7Den.pdf

[43] J. Duron, F. Grilli, B. Dutoit *et al.*, "Modelling the E–J relation of high-Tc superconductors in an arbitrary current range," *Physica C*, vol. 401, iss. 1-4, pp. 231-235, Jan. 2004.

[44] J. Duron, F. Grilli, L. Antognazza *et al.*, "Finite-element modelling of superconductors in over-critical regime with temperature dependent resistivity," *J. Phys. Conf. Ser.*, vol. 43, pp. 1076-1080, 2006.

[45] M. D. Ainslie, D. Zhou, H. Fujishiro *et al.*, "Flux jump-assisted pulsed field magnetization of high-Jc bulk high temperature superconductors", *Supercond. Sci. Technol.*, vol. 29, no. 12, pp. 124004, Oct. 2016.

[46] F. Fiorillo, "Measurement and characterization of magnetic materials," Elsevier Academic Press, 2004, ISBN 0-12-257251-3, pp. 31.

[47] J. Li, N. Qian, S. Si *et al.*, "Vibration characteristics of the HTS maglev vehicle running on a 45-m-long ring test line", *IEEE Trans. Appl. Supercond.*, vol. 26, no. 6, pp. 3602507, Jun. 2016.

[48] W. Yang, M. Qiu, Y. Liu *et al.*, "Levitation characteristics in an HTS maglev launch assist test vehicle," *Supercond. Sci. Technol.*, vol. 20, no. 3, pp. 281-286, Feb. 2007.

[49] A. Patel, S. C. Hopkins, A. Baskys *et al.*, "Magnetic levitation using high temperature superconducting pancake coils as composite bulk cylinders," *Supercond. Sci. Technol.*, vol. 28, no. 11, pp. 115007, Sep. 2015.

[50] F. N. Werfel, U. Floegel-Delor, T. Riedel *et al.*, "HTS magnetic bearings in prototype application", *IEEE Trans. Appl. Supercond.*, vol. 20, no. 3, pp. 874-879, Jun. 2010.

[51] A. C. Day, M. Strasik, K. E. McCrary *et al.*, "Design and testing of the HTS bearing for a 10 kWh flywheel system," *Supercond. Sci. Technol.*, vol. 15, pp. 838-841, Apr. 2002.

[52] D. Zhou, M. Izumi, M. Miki *et al.*, "An overview of rotating machine systems with high temperature bulk superconductors," *Supercond. Sci. Technol.*, vol. 25, no. 10, pp. 103001, Aug. 2012.

[53] T. Nakamura, D. Tamada, Y. Yanagi *et al.*, "Development of a superconducting bulk magnet for NMR and MRI," *J. Magn. Reson.*, vol. 259, pp. 68-75, Oct. 2015.

[54] S. B. Kim, M. Imai, R. Takano *et al.*, "Study on optimized configuration of stacked HTS bulk annuli for compact NMR application", *IEEE Trans. Appl. Supercond.*, vol. 21, no. 3, pp. 2080-2083, Jun. 2011.

[55] T. Oka, H. Kanayama, S. Fukui *et al.*, "Application of HTS bulk magnet system to the magnetic separation techniques for water purification," *Physica C*, vol. 468, iss. 15-20, pp. 2128-2132, Sep, 2008.

[56] N. Nishijima, N. Saho, K. Asano *et al.*, “Magnetization method for long high-T_c bulk superconductors used for magnetic separation”, *IEEE Trans. Appl. Supercond.*, vol. 13, no. 2, pp. 1580-1583, Jun. 2003.

[57] A. M. Campbell and D. A. Cardwell, “Bulk high temperature superconductors for magnet applications,” *Cryogenics*, vol. 37, iss. 10, pp. 567-575, 1997.

[58] A. Patel, A. Baskys, S. C. Hopkins, “Pulsed-field magnetization of superconducting tape stacks for motor applications”, *IEEE Trans. Appl. Supercond.*, vol. 25, no. 3, pp. 5203405, Jun. 2015.

[59] D. Ishizuka, S.B. Kim, N. Hayashi *et al.*, "Numerical study on split coil-shaped HTS bulks to improve the field homogeneity for compact NMR relaxometry magnets", *Phys. Procedia*, vol. 58, pp. 290-293, 2014.

[60] K. Hojo, S.B. Kim, D. Miyazawa *et al.*, "Study on the magnetic field homogeneity of stacked HTS bulk magnets Including the deteriorated HTS bulk by crack for compact NMR relaxometry", *Phys. Procedia*, vol. 81, pp. 166-169, 2016

[61] T. A. Coombs, Z. Hong, X. Zhu *et al.*, “A novel heat engine for magnetizing superconductors”, *Supercond. Sci. Technol.*, vol. 21, no. 3, pp. 034001, Feb. 2008.

[62] C. Hoffmann, D. Pooke and A. D. Caplin, “Flux pump for HTS magnets,” *IEEE Trans. Appl. Supercond.*, vol. 21, no. 3, pp. 1628-1631, Jun. 2011.

[63] J. Geng, K. Matsuda, L. Fu *et al.*, “Origin of dc voltage in type II superconducting flux pumps: field, field rate of change, and current density dependence of resistivity,” *Supercond. Sci. Technol.*, vol. 49, no. 11, pp. 11LT01, Feb. 2016.

[64] M. Tomita and M. Murakami, “High temperature superconductor bulk magnets that can trap magnetic fields of over 17 tesla at 29 K,” *Nature*, vol. 421, pp. 517-520, Jan. 2003.

[65] S. Nariki, N. Sakai and M. Murakami, “Melt-processed Gd–Ba–Cu–O superconductor with trapped field of 3 T at 77 K,” *Supercond. Sci. Technol.*, vol. 18, no. 2, pp. S126-S130, Dec. 2005.

[66] G. Fuchs, W. Häßler, K. Nenkov *et al.*, “High trapped fields in bulk MgB_2 prepared by hot-pressing of ball-milled precursor powder,” *Supercond. Sci. Technol.*, vol. 26, no. 12, pp. 122002, Nov. 2013.

[67] T. Tamegai, T. Hirai, Y. Sun *et al.*, "Trapping a magnetic field of 7.9 T using a bulk magnet fabricated from stack of coated conductors," *Physica C*, vol. 530, pp. 20-23, Nov. 2016.

[68] A. Patel, K. Filar, V. I. Nizhankovskii *et al.*, "Trapped fields greater than 7 T in a 12 mm square stack of commercial high temperature superconducting tape," *Appl. Phys. Lett.*, vol. 102, iss. 10, pp. 102601, Feb. 2013.

[69] U. Mizutani, T. Oka, Y. Itoh *et al.*, "Pulsed-field magnetization applied to high-Tc superconductors," *Appl. Supercond.*, vol. 6, iss. 2-5, pp. 235-246, Feb.-May 1998.

[70] H. Fujishiro, T. Oka, K. Yokoyama *et al.*, "Time evolution and spatial distribution of temperature in YBCO bulk superconductor after pulse field magnetizing," *Supercond. Sci. Technol.*, vol. 16, no. 7, pp. 809-814, Jun. 2003.

[71] H. Fujishiro, M. Kaneyama, K. Yokoyama *et al.*, "Generated heat during pulse field magnetizing for REBaCuO (RE = Gd, Sm, Y) bulk superconductors with different pinning abilities," *Supercond. Sci. Technol.*, vol. 18, no. 1, pp. 158-165, Nov. 2004.

[72] H. Fujishiro, T. Tateiwa, A. Fujiwara *et al.*, "Higher trapped field over 5 T on HTSC bulk by modified pulse field magnetizing," *Physica C*, vol. 445-448, no. 1, pp. 334-338, Oct. 2006.

[73] A. Patel, S. C. Hopkins and B. A. Glowacki, "Trapped fields up to 2 T in a 12 mm square stack of commercial superconducting tape using pulsed field magnetization," *Supercond. Sci. Technol.*, vol. 26, no. 3, pp. 032001, Jan. 2013.

[74] T. Ida, H. Matsuzaki, Y. Akita *et al.*, "Magnetization properties for Gd–Ba–Cu–O bulk superconductors with a couple of pulsed-field vortex-type," *Physica C*, vol. 412-414, pp. 638-645, 2004.

[75] H. Fujishiro, T. Naito, and M. Oyama, "Mechanism of magnetic flux trapping on superconducting bulk magnetized by pulsed field using a vortex type coil," *Supercond. Sci. Technol.*, vol. 24, no. 7, pp. 075015, Jun. 2011

[76] H. Fujishiro, M. Ikebe, T. Naito *et al.*, "Anisotropic thermal diffusivity and conductivity of YBCO(123) and YBCO(211) mixed crystals. I," *Jpn. J. Appl. Phys.*, vol. 33, part 1, no. 9A, pp. 4965-4970, Sep. 1994.

[77] Y. Kimura, K. Yamaguchi, T. Sano *et al.,* "Practical technique of pulsed field magnetization for bulk HTS application," *J. Phys. Conf. Ser.*, vol. 97, pp. 012295, 2008.

[78] H. Fujishiro, T. Naito, K. Kakehata *et al.,* "Estimation of temperature rise from trapped field gradient on superconducting bulk magnetized by multi-pulse technique," *Supercond. Sci. Technol.*, vol. 23, no. 2, pp. 025013, Dec. 2009.

[79] H. Fujishiro, T. Hiyama, T. Naito *et al.,* "Enhancement of trapped field and total trapped flux on GdBaCuO bulk by the MMPSC+IMRA method," *Supercond. Sci. Technol.*, vol. 22, no. 9, pp. 095006, Aug. 2009.

[80] M. Sander, U. Sutter, R. Koch *et al.,* "Pulsed magnetization of HTS bulk parts at T<77 K," *Supercond. Sci. Technol.*, vol. 13, no. 6, 2000.

[81] H. Kamijo and H. Fujimoto, "Repeated pulsed-field magnetization with temperature control in a high-T_c bulk superconductor," *IEEE Trans. Appl. Supercond.*, vol. 11, no. 1, pp. 1816-1819, Mar. 2001.

[82] K. Kajikawa, R. Yokoo, K. Tomachi *et al.*, "Finite element analysis of pulsed field magnetization process in a cylindrical bulk superconductor," *Physica C*, vol. 468, iss. 15-20, pp. 1494-1497, Sep. 2008.

[83] H. Fujishiro, T. Naito and M. Oyama, "Simulation of flux dynamics in a superconducting bulk magnetized by multi-pulse technique," *Physica C*, vol. 471, iss. 21-22, pp. 889-892, Nov. 2011.

[84] A. G. Page, A. Patel, A. Baskys *et al.*, "The effect of stabilizer on the trapped field of stacks of superconducting tape magnetized by a pulsed field," *Supercond. Sci. Technol.*, vol. 28, no. 8, pp. 085009, Jul. 2015.

[85] A. Patel and B. A. Glowacki, "Enhanced trapped field achieved in a superconducting bulk using high thermal conductivity structures following simulated pulsed field magnetization," *Supercond. Sci. Technol.*, vol. 25, no. 12, pp. 125015, Nov. 2012.

[86] A. Patel and B. Glowacki, "Optimisation of composite superconducting bulks made from (RE)BCO coated conductor stacks using pulsed field magnetization modelling," *J. Phys. Conf. Ser.*, vol. 507, no. 2, pp. 022024, 2014.

[87] K. Yokoyama, T. Oka, N. Kondo *et al.*, “Evaluation of pulsed-field magnetization of a bulk superconductor with small holes,” *IEEE Trans. Appl. Supercond.*, vol. 23, no. 3, pp. 8201204, Jun. 2013.

[88] E. H. Brandt, "Square and rectangular thin superconductors in a transverse magnetic field", *Phy. Rev. Lett.*, vol. 74, no. 15, pp. 3025-3028, Apr. 1995.

[89] E. H. Brandt, "Superconductors of finite thickness in a perpendicular magnetic field: Strips and slabs", *Phy. Rev. B*, vol. 54, no. 6, pp. 4246-4264, Aug. 1996.

[90] E. H. Brandt, "Superconductor disks and cylinders in an axial magnetic field. I. Flux penetration and magnetization curves", *Phy. Rev. B*, vol. 58, no. 10, pp. 6506-6522, Sep. 1998.

[91] E. H. Brandt, “Thin superconductors in a perpendicular magnetic AC field. II. Circular disk,” *Phys. Rev. B*, vol. 50, no. 6, pp. 4034-4050, Aug. 1994.

[92] E. H. Brandt, M. Indenbom, “Type-II-superconductor strip with current in a perpendicular magnetic field,” *Phys. Rev. B*, vol. 48, no. 17, pp. 12893-12906, Nov. 1993.

[93] E. Zeldov, J. R. Clem, M. McElfresh *et al.* "Magnetization and transport currents in thin superconducting films", *Phy. Rev. B*, vol. 49, no. 14, pp. 9802-9822, Apr. 1994.

[94] R. Hiptmair, “Finite elements in computational electromagnetism,” pp. 237-339, Cambridge University Press, 2002.

[95] P. Krüger, “Optimisation of hysteresis losses in high-temperature superconducting wires,” KIT Scientific Publishing, ISSN 1869-1765, 2013.

[96] W. T. Norris, “Calculation of hysteresis losses in hard superconductors carrying ac: isolated conductors and edges of thin sheets,” *J. Phys. D: Appl. Phys.*, vol. 3, no. 4, pp. 489-507, 1970.

[97] J. R. Clem and A. Sanchez, “Hysteretic AC losses and susceptibility of thin superconducting disks,” *Phys. Rev. B*, vol. 50, no. 13, pp. 9355-9362, Oct. 1994.

[98] E. Pardo, D. X. Chen, A. Sanchez *et al.*, “The transverse critical-state susceptibility of rectangular bars,” *Supercond. Sci. Technol.*, vol. 17, no. 3, pp. 537-544, Feb. 2004.

[99] M. R. Halse, “AC face field losses in a type II superconductor,” *J. Phys. D: Appl. Phys.*, vol. 3, no. 5, pp. 717-720, 1970.

[100] G. P. Mikitik, Y. Mawatari, A. T. S. Wan *et al.*, “Analytical methods and formulas for modeling high temperature superconductors,” *IEEE Trans. Appl. Supercond.*, vol. 23, no. 2, pp. 8001920, Apr. 2013.

[101] S. Stavrev, F. Grilli, B. Dutoit *et al.*, “Comparison of numerical methods for modeling of superconductors,” *IEEE Trans. Appl. Supercond.*, vol. 38, no. 2, pp. 849-852, Mar. 2002.

[102] O. Tsukamoto, N. Sekine, M. Ciszek *et al.*, “A method to reduce magnetization losses in assembled conductors made of YBCO coated conductors,” *IEEE Trans. Appl. Supercond.*, vol. 15, no. 2, pp. 2823-2826, Jun. 2005.

[103] K. Seo, K. Fukuhara, and M. Hasegawa, “Analyses for inter-strand coupling loss in multi-strand superconducting cable with distributed contact resistance between strands,” *Cryogenics*, vol. 412, iss. 2, pp. 131-137, Feb. 2001.

[104] X. Wang, S. Hahn, Y. Kim *et al.*, “Turn-to-turn contact characteristics for an equivalent circuit model of no-insulation ReBCO pancake coil,” *Supercond. Sci. Technol.*, vol. 26, no. 3, pp. 035012, Jan. 2013.

[105] A. M. Campbell, “An Introduction to numerical methods in superconductors,” *J. Supercond. Nov. Magn.,* vol 24, iss. 1-2, pp. 27-33, Jan. 2011.

[106] F. Grilli, “Numerical modeling of HTS applications,” *IEEE Trans. Appl. Supercond.*, vol. 26, no. 3, pp. 0500408, Apr. 2016.

[107] A. Kameni, J. Lambrechts, J. F. Remacle *et al.*, “Discontinuous Galerkin method for computing induced fields in superconducting materials,” *IEEE Trans. Appl. Supercond.*, vol. 48, no. 2, pp. 591-594, Feb. 2012.

[108] A. Kameni, M. Boubekeur, L. Alloui *et al.*, “A 3-D semi-implicit method for computing the current density in bulk superconductors,” *IEEE Trans. Appl. Supercond.*, vol. 50, no. 2, pp. 7009204, Feb. 2014.

[109] A. Badia and C. Lopez, "Vector magnetic hysteresis of hard superconductors", *Phy. Rev. B*, vol. 65, pp. 104514, Feb. 2002.

[110] K. Kajikawa, T. Hayashi, R. Yoshida *et al.*, "Numerical evaluation of AC losses in HTS wires with 2D FEM formulated by self-magnetic field," *IEEE Trans. Appl. Supercond.*, vol. 13, no. 2, pp. 3630-3633, Jun. 2003.

[111] R. Brambilla, F. Grilli and L. Martini, "Development of an edge-element model for AC loss computation of high-temperature superconductors," *Supercond. Sci. Technol.*, vol. 20, no. 1, pp. 16-24, Nov. 2006.

[112] https://www.comsol.com

[113] http://www.htsmodelling.com/?page_id=748, model file 6 by M. D. Ainslie

[114] F. Grilli, V. M. R. Zermeño, M. Vojenčiak *et al.*, "AC losses of pancake coils made of Roebel cable," *IEEE Trans. Appl. Supercond.*, vol. 23, no. 3, pp. 5900205, Jun. 2013.

[115] http://i-sunam.com

[116] L. Ying, J. Xu, J. Sheng *et al.*, "Numerical and experimental analysis of AC Loss of YBCO coated conductor carrying DC and AC offset," *IEEE Trans. Appl. Supercond.*, vol. 23, no. 3, pp. 5900704, Jun. 2013

[117] A. F. Zhao, X. F. Li, W. Wu *et al.*, "AC Loss Characteristics of HTS wires carrying currents with different waveforms," *IEEE Trans. Appl. Supercond.*, vol. 26, no. 4, pp. 8201805, Jun. 2016.

[118] O. Tsukamoto, "AC losses in a type II superconductor strip with inhomogeneous critical current distribution," *Supercond. Sci. Technol.*, vol. 18, no. 5, pp. 596-605, Mar. 2005.

[119] T. Nishioka, N. Amemiya, N. Enomoto *et al.*, "AC loss of YBCO coated conductors fabricated by IBAD/PLD method," *IEEE Trans. Appl. Supercond.*, vol. 15, no. 2, pp. 2843-2846, Jun. 2005.

[120] M. Solovyov, E. Pardo, J. Šouc *et,al.,* "Non-uniformity of coated conductor tapes," *Supercond. Sci. Technol.*, vol. 26, no. 11, pp. 115013, Oct. 2013.

[121] F. Liang, W. Yuan, M. Zhang *et al.*, "AC loss modelling and experiment of two types of low-inductance solenoidal coils," *Supercond. Sci. Technol.*, vol. 29, no. 11, pp. 115006, Oct. 2016.

[122] J. Šouc, F. Gömöry and M. Vojenčiak, "Calibration free method for measurement of the AC magnetization loss," *Supercond. Sci. Technol.*, vol. 18, no. 5, pp. 592-595, Mar. 2005.

[123] F. Grilli, R. Brambilla and L. Martini, "Modeling high-temperature superconducting tapes by means of edge finite elements," *IEEE Trans. Appl. Supercond.*, vol. 17, no. 2, pp. 3155-3158, Jun. 2007.

[124] Yunus A. Cengel, "Heat Transfer: A Practical Approach [2nd Edition]," Mcgraw-Hill, Oct, 2002.

[125] R. Shiraishi and H. Ohsaki, "Flux dynamics in inhomogeneous bulk superconductor during pulsed field magnetization," *IEEE Trans. Appl. Supercond.*, vol. 16, no. 2, pp. 1794-1797, Jun. 2006.

[126] Y. Komi, M. Sekino and H. Ohsaki, "Three-dimensional numerical analysis of magnetic and thermal fields during pulsed field magnetization of bulk superconductors with inhomogeneous superconducting properties," *Physica C*, vol. 469, iss. 15-20, pp. 1262-1265, Oct. 2009.

[127] K. Berger, J. Lévêque, D. Netter *et al.*, "Influence of temperature and/or field dependences of the E–J power law on trapped magnetic field in bulk YBaCuO," *IEEE Trans. Appl. Supercond.*, vol. 17, no. 2, pp. 3028-3031, Jun. 2007.

[128] H. Fujishiro and T. Naito, "Simulation of temperature and magnetic field distribution in superconducting bulk during pulsed field magnetization," *Supercond. Sci. Technol.*, vol. 23, no. 10, pp. 105021, Sep. 2010.

[129] Z. Xu, R. Lewin, A. M. Campbell *et al.*, "Simulation studies on the magnetization of (RE)BCO bulk superconductors using various split-coil arrangements," *Supercond. Sci. Technol.*, vol. 25, no. 2, pp. 025016, Jan. 2012.

[130] J. Zou, M. D. Ainslie, D. Hu *et al.*, "Influence of time-varying external magnetic Fields on trapped fields in bulk superconductors," *IEEE Trans. Appl. Supercond.*, vol. 25, no. 3, pp. 4900505, Jun. 2015.

[131] T. Matsushita, E. S. Otabe, T. Fukunaga *et al.*, "Week link property in superconducting Y-Ba-Cu-O prepared by QMG process," *IEEE Trans. Appl. Supercond.*, vol. 3, no. 1, pp. 1045-1048, Mar. 1993.

[132] H. Fukai, M. Tomita, M. Murakami *et al.*, “The effect of geometry on the trapped magnetic field in bulk superconductors,” *Supercond. Sci. Technol.*, vol. 15, no. 7, pp. 1054-1057, May 2002.

[133] M. P. Philippe, J. F. Fagnard, S. Kirsch *et al.*, “Magnetic characterization of large grain, bulk Y–Ba–Cu–O superconductor–soft ferromagnetic alloy hybrid structures,” *Physica C*, vol. 502, pp. 20-30, Jul. 2014.

[134] N. Schönborg and S. Hörnfeldt, “Model of the temperature dependence of the hysteresis losses in a high-temperature superconductor,” *Physica C*, vol. 372-376, part 3, pp. 1734-1738, Aug. 2002.

[135] Y. Yeshurun, A. P. Malozemoff and A. Shaulov, “Magnetic relaxation in high-temperature superconductors”, *Rev. Mod. Phys.*, vol 68, no. 3, pp. 911-949, Jul. 1996.

[136] D. A. Cardwell, M. Murakami, M. Zeisberger *et al.*, “Round robin measurements of the flux trapping properties of melt processed Sm–Ba–Cu–O bulk superconductors,” *Physica C*, vol. part 1, 414-414, pp. 623-632, Oct. 2004.

[137] A. Gurevich and H. Küpfer, “Time scales of the flux creep in superconductors,” *Phys. Rev. B.*, vo. 48, no. 9, pp. 6477-6487, Sep. 1993.

[138] J. López, R. Maynou, X. Granados *et al.*, “Finite element analysis of thermal and electromagnetic relaxation processes in a YBCO currents distributions differences,” *IEEE Trans. Appl. Supercond.*, vol. 23, no. 3, pp. 4901104, Jun. 2013.

[139] J. López, R. Maynou, X. Granados *et al.*, “Analysis of magnetic remanence calculated in HTS by pulse magnetization process taking into account the temperature effect,” *Phys. Procedia.*, vol. 36, pp. 1189-1194, 2012.

[140] V. M. R. Zermeño *et al.*, “Estimation of maximum possible trapped field in superconducting magnets in 2D and 3D,” presented at the *Int. Conf. on Magnet Technology 24* (Presentation ID: 2PoBE_02).

[141] A. L. Rakhmanov, V. S. Vysotsky, Y. A. Ilyin *et al.*, “Universal scaling law for quench development in HTSC devices,” *Cryogenics*, vol. 40, iss. 1, pp. 19-27, Jan. 2000.

[142] M. Lindmayer and H. Mosebach, "Current limiting properties of YBCO-films on sapphire substrates," *IEEE Trans. Appl. Supercond.*, vol. 9, no. 2, pp. 1369-1372, Jun. 1999.

[143] J. Duron, L. Antognazza, M. Decroux *et al.*, "3-D finite element simulations of strip lines in a YBCO/Au fault current limiter," *IEEE Trans. Appl. Supercond.*, vol. 15, no. 2, pp. 1998-2002, Jun. 2005.

[144] A. Baskys, A. Patel, S. C. Hopkins *et al.*, "Modeling of trapped fields by stacked (RE)BCO tape using angular transversal field dependence," *IEEE Trans. Appl. Supercond.*, vol. 26, no. 3, pp. 6601004, Apr. 2016.

[145] S. Zou, V. M. R. Zermeño and F. Grilli, "Influence of parameters on the simulation of HTS bulks magnetized by pulsed field magnetization," *IEEE Trans. Appl. Supercond.*, vol. 26, no. 4, pp. 4702405, Jun. 2016.

[146] Y. Yanagi, Y. Itoh, M. Yoshikawa *et al.*, "Pulsed field magnetization of a 36 mm diameter single-domain Sm–Ba–Cu–O bulk superconductor at 30, 35 and 77 K," *Supercond. Sci. Technol.*, vol. 18, no. 6, pp. 839-849, Apr. 2005.

[147] Y. Itoh and U. Mizutani, "Pulsed field magnetization of melt-processed Y-Ba-Cu-O superconducting bulk magnet," *Jpn. J. Appl. Phys.*, vol. 35, Part 1, no. 4A, pp. 2114-2125, Apr. 1996.

[148] H. Fujishiro, T. Hiyama, T. Miura *et al.*, "Pulsed field magnetization for GdBaCuO bulk with stronger pinning characteristics," *IEEE Trans. Appl. Supercond.*, vol. 19, no. 3, pp. 3545-3548, Jun. 2009.

[149] Y. Yanagi, Y. Itoh, M. Yoshikawa *et al.*, "Pulsed-field magnetization of Y-Ba-Cu-O superconducting bulk magnet cooled by refrigerator," *Adv. Supercond. X. Springer*, Tokyo, pp. 941-944, 1998.

[150] "Status of 2G HTS Wire for Electric Power Applications," (2013). http://www.superpower-inc.com/system/files/2013_0426+CIGRE+WG38+Wksp_SuperPower.pdf

[151] J. Lu, E. S. Choi and H. D. Zhou, "Physical properties of Hastelloy® C-276™ at cryogenic temperatures," *J. Appl. Phys.*, vol. 103, no. 6, pp. 064908, Mar. 2008.

[152] M. Bonura and C. Senatore, “High-field thermal transport properties of REBCO coated conductors,” *Supercond. Sci. Technol.*, vol. 28, no. 2, pp. 025001, Dec. 2014.

[153] G. Ahlers, “Heat capacity of Copper,” *Rev. Sci. Instrum.*, vol. 37, iss. 4 pp. 477-480, Dec. 1965.

[154] R. A. Matula, “Electrical resistivity of copper, gold, palladium, and silver,” *J. Phys. Chem. Ref. Data*, vol. 8, no. 4, pp. 1147-1298, 1979.

[155] D. R. Smith and F. R. Fickett, “Low-temperature properties of silver,” *J. Res. Natl. Inst. Stand. Technol.*, vol. 100, no. 2, pp. 119-171, 1995.

[156] http://www.htsmodelling.com/?page_id=1039

[157] S. Zou, V. M. R. Zermeño and F. Grilli, “Simulation of stacks of high-temperature superconducting coated conductors magnetized by pulsed field magnetization using controlled magnetic density distribution coils,” *IEEE Trans. Appl. Supercond.*, vol. 26, no. 3, pp. 8200705, Apr. 2016.

[158] S. Zou, V. M. R. Zermeño, A. Baskys *et,al.*, “Simulation and experiments of stacks of high temperature superconducting coated conductors magnetized by pulsed field magnetization with multi-pulse technique,” *Supercond. Sci. Technol.*, vol. 30, no. 1, pp. 014010, Nov. 2016.

[159] F. Grilli, R. Brambilla, F. Sirioos *et al.*, “Development of a three-dimensional finite-element model for high-temperature superconductors based on the H-formulation,” *Croygenics*, vol. 53, pp. 142-147, Jan. 2013.

[160] A. Stenvall, F. Grilli and M. Lyly, “Current-penetration patterns in twisted superconductors in self-field,” *IEEE Trans. Appl. Supercond.*, vol. 23, no. 3, pp. 8200105, Jun. 2013.

[161] V. M. R. Zermeño and F. Grilli, “3D modeling and simulation of 2G HTS stacks and coils,” *Supercond. Sci. Technol.*, vol. 27, no. 4, pp. 044025, Mar. 2014.

[162] V. M. R. Zermeño, F. Grilli and F. Sirois, “A full 3D time-dependent electromagnetic model for Roebel cables,” *Supercond. Sci. Technol.*, vol. 26, no. 5, pp. 052001, Mar. 2013.

[163] S. Lee, V. Petrykin, A. Molodyk *et al.*, "Development and production of second generation high Tc superconducting tapes at SuperOx and first tests of model cables," *Supercond. Sci. Technol.*, vol. 27, no. 4, pp. 044022, Mar. 2014.

[164] http://www.superox.ru/en/products/42-2G-HTS-tape/

[165] Ch. Jooss, J. Albrecht, H. Kuhn *et al.*, "Magneto-optical studies of current distributions in high-Tc superconductors," *Rep. Prog. Phys*, vol. 65, pp. 651-788, Apr. 2002.

[166] http://www.arepoc.sk

Karlsruher Schriftenreihe zur Supraleitung (ISSN 1869-1765)

Herausgeber: Prof. Dr.-Ing. M. Noe, Prof. Dr. rer. nat. M. Siegel

Die Bände sind unter www.ksp.kit.edu als PDF frei verfügbar oder als Druckausgabe bestellbar.

Band 001
Christian Schacherer
Theoretische und experimentelle Untersuchungen zur Entwicklung supraleitender resistiver Strombegrenzer. 2009
ISBN 978-3-86644-412-6

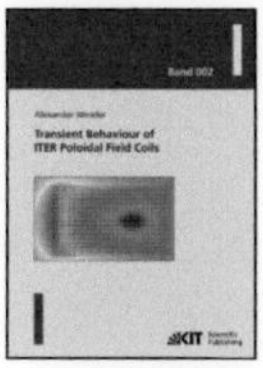

Band 002
Alexander Winkler
Transient behaviour of ITER poloidal field coils. 2011
ISBN 978-3-86644-595-6

Band 003
André Berger
Entwicklung supraleitender, strombegrenzender Transformatoren. 2011
ISBN 978-3-86644-637-3

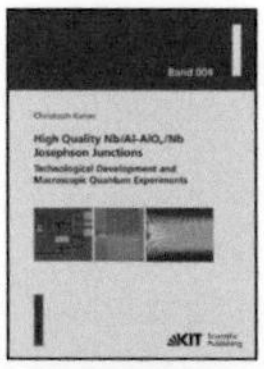

Band 004
Christoph Kaiser
High quality Nb/Al-AlOx/Nb Josephson junctions. Technological development and macroscopic quantum experiments. 2011
ISBN 978-3-86644-651-9

Band 005
Gerd Hammer
Untersuchung der Eigenschaften von planaren Mikrowellenresonatoren für Kinetic-Inductance Detektoren bei 4,2 K. 2011
ISBN 978-3-86644-715-8

Die Bände sind unter www.ksp.kit.edu als PDF frei verfügbar oder als Druckausgabe bestellbar.

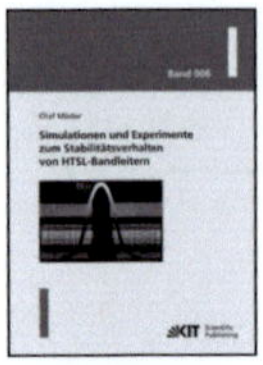

Band 006
Olaf Mäder
Simulationen und Experimente zum Stabilitätsverhalten von HTSL-Bandleitern. 2012
ISBN 978-3-86644-868-1

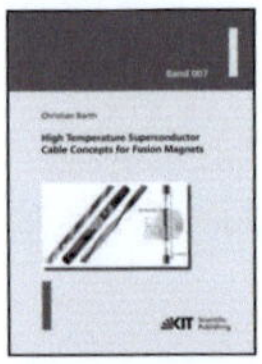

Band 007
Christian Barth
High Temperature Superconductor Cable Concepts for Fusion Magnets. 2013
ISBN 978-3-7315-0065-0

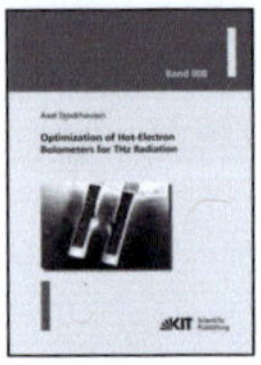

Band 008
Axel Stockhausen
Optimization of Hot-Electron Bolometers for THz Radiation. 2013
ISBN 978-3-7315-0066-7

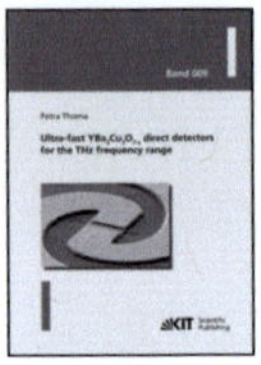

Band 009
Petra Thoma
Ultra-fast $YBa_2Cu_3O_{7-x}$ direct detectors for the THz frequency range. 2013
ISBN 978-3-7315-0070-4

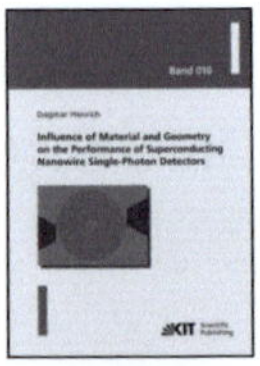

Band 010
Dagmar Henrich
Influence of Material and Geometry on the Performance of Superconducting Nanowire Single-Photon Detectors. 2013
ISBN 978-3-7315-0092-6

Band 011
Alexander Scheuring
Ultrabreitbandige Strahlungseinkopplung in THz-Detektoren. 2013
ISBN 978-3-7315-0102-2

Die Bände sind unter www.ksp.kit.edu als PDF frei verfügbar oder als Druckausgabe bestellbar.

Band 012
Markus Rösch
Development of lumped element kinetic inductance detectors for mm-wave astronomy at the IRAM 30 m telescope. 2013
ISBN 978-3-7315-0110-7

Band 013
Johannes Maximilian Meckbach
Superconducting Multilayer Technology for Josephson Devices. 2013
ISBN 978-3-7315-0122-0

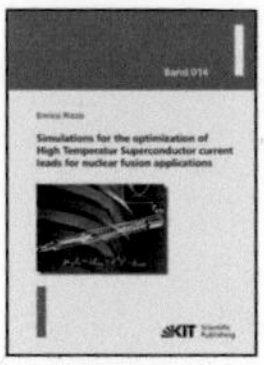

Band 014
Enrico Rizzo
Simulations for the optimization of High Temperature Superconductor current leads for nuclear fusion applications. 2014
ISBN 978-3-7315-0132-9

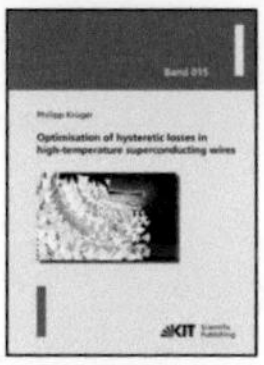

Band 015
Philipp Krüger
Optimisation of hysteretic losses in high-temperature superconducting wires. 2014
ISBN 978-3-7315-0185-5

Band 016
Matthias Hofherr
Real-time imaging systems for superconducting nanowire single-photon detector arrays. 2014
ISBN 978-3-7315-0229-6

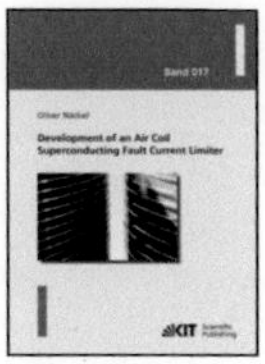

Band 017
Oliver Näckel
Development of an Air Coil Superconducting Fault Current Limiter. 2016
ISBN 978-3-7315-0526-6

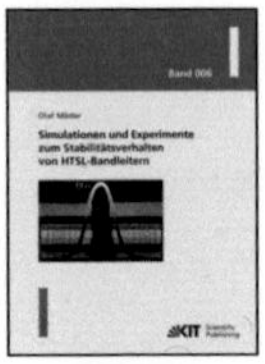

Band 018
Christoph M. Bayer
Characterization of High Temperature Superconductor Cables for Magnet Toroidal Field Coils of the DEMO Fusion Power Plant. 2017
ISBN 978-3-7315-0605-8

Band 019
Shengnan Zou
Magnetization of High Temperature Superconducting Trapped-Field Magnets. 2017
ISBN 978-3-7315-0715-4